# IdeaFest Journal

2024 -2025

The Press at Cal Poly Humboldt

*T*HE PRESS AT CAL POLY HUMBOLDT
Cal Poly Humboldt Library
1 Harpst St.
Arcata, CA 95524-8299
digitalcommons.humboldt.edu


ISBN: 978-1-962081-24-5
ISSN: 2689-1891

DESIGN AND LAYOUT: SARAH GODLIN
ADDITIONAL EDITING, AARON LAUGHLIN

The Press at Cal Poly Humboldt publishes high-quality scholarly, intellectual, and creative works by or in support of our campus community. The press supports the Cal Poly Humboldt mission to improve the human condition and our environment by promoting understanding of social, economic, and environmental issues.

# Contents

IdeaFest
Journal

# Implications of Using eDNA as a Monitoring Tool in the Elk River Estuary

**Christy Wheatley, Fernando Martinez, Trinity Spencer**

## Abstract

This study compares the efficacy of environmental DNA (eDNA) sampling to traditional capture-based methods such as seine netting and minnow trapping for monitoring fish assemblages in the Elk River estuary. The Elk River estuary is located in Humboldt County, California and hosts a diverse range of fish species due to its tidal influences and brackish habitat. A restoration is planned for this area, thus it is important to explore the effectiveness of different fish monitoring methods that could be used to evaluate restoration success. For this capstone project, we conducted a pilot study at five sites within the Elk River estuary, where water samples for eDNA analysis of fish species were collected alongside capture-based fish sampling methods (seining and minnow trapping) for comparison. These five sites are potentially inhabited by several species of interest including: coho salmon (*Oncorhynchus kisutch*), Chinook salmon (*Oncorhynchus tshawytscha*), steelhead trout (*Oncorhynchus mykiss*), coastal cutthroat trout (*Oncorhynchus clarkii clarkii*), tidewater goby (*Eucyclogobius newberryi*), and longfin smelt (*Spirinchus thaleichthys*). We then generated a monitoring plan based on our findings from the pilot study with sites for immediate pre-restoration sampling and sites where sampling should be conducted once the restoration project is completed. Included in the monitoring plan are limitations to eDNA sampling with reflections on the strengths of both eDNA and capture-based samplings. Overall, we found that eDNA sampling was less labor intensive and easier to conduct compared to seining and minnow trapping. The results from this pilot study and subsequent monitoring plan are intended to help organizations such as California Trout (or CalTrout) to determine the most effective and cost-efficient methods to monitor fish species as part of restoration post-project assessment.

# 1. Introduction

## 1.1 Importance of fisheries monitoring

Monitoring is a key aspect of restoration projects because it can evaluate the effectiveness of the restoration actions. A study conducted by Bennet et al. (2016) found that intensive monitoring is the best approach for understanding the impacts of restoration. Despite the recognized importance of restoration project monitoring, in practice, there is often a lack of sufficient post-project assessment for most restoration projects (Roni et al., 2008). In fact, there is a lack of evidence supporting most restoration actions, meaning that conclusions cannot be drawn about best practices (Roni et al., 2008).

Fish populations nationwide are struggling due to overfishing, warming waters, pollution, and a decline in biodiversity. Fish have high importance in the environment, economy, and communities that have relied on them for centuries (Finegold, 2009). The loss of biodiversity is devastating for aquatic ecosystems, because the variety of species present in each respective waterway reflects the overall health of that ecosystem (Moyle & Leidy, 1992). Changes in fish biodiversity affect the structure and function of marine ecosystems, leading to depleted resources and the extinction of many species (Hiddink et al., 2007). In Washington, salmon populations and watershed health have been declining, resulting in more stream and river restoration projects (Bash & Ryan, 2002). Project managers who have received a "hydraulic project approval" were surveyed to determine if monitoring was part of their restoration project (Bash & Ryan, 2002). Half of the managers surveyed reported taking baseline data; of those half, only 18% said monitoring was required for their projects (Bash & Ryan, 2002). The authors state that monitoring is a voluntary process, and monitoring efforts ultimately depend on funding, stakeholders, and managers (Bash & Ryan, 2002).

## 1.2 eDNA sampling

Environmental DNA (eDNA) is a monitoring tool that utilizes the DNA left in the environment by terrestrial and aquatic species to determine biodiversity (Beng & Corlett, 2020). eDNA can be found in soil, sediment, water or snow, although sampling is best used in aquatic ecosystems because the DNA is suspended in the water column, allowing any part of the waterbody to be sampled and yield results (Rees et al., 2014). The DNA is extracted from water samples, processed using a polymerase chain reaction (or PCR), and sequenced to generate genetic data (Rupert et al., 2019). Because of the ease of use in water bodies, eDNA sampling is being used more frequently by fisheries managers to detect the presence of threatened, endangered, and invasive fish (Thomas et al., 2018). While this emerging technology can effectively monitor fish population biodiversity, questions remain about its best usage and application (Ruppert et al., 2019).

One study of eDNA sampling found that effectiveness is limited by certain water quality conditions, including suspended sediment, pollution, and tannins (Joseph et al., 2022). These limitations can make it difficult to obtain reliable results in highly polluted or turbid waterways (DiBattista, 2022). DNA is also altered by UV radiation, temperature changes, chemicals (such as insecticides), and other biotic and abiotic factors, which can potentially lead to false negative results (Joseph et al., 2022). Despite these considerations, eDNA is still very effective. A study conducted by Smart et al. (2015) compared bottle trapping and eDNA sampling on *Lissotriton vulgaris vulgaris,* an invasive smooth newt that has been detected for the first time in Melbourne, Australia. The authors of this study found eDNA sampling to be much more sensitive than other sampling methods (in this case, bottle trapping) with per-sample eDNA detection probabilities being a full magnitude greater than bottle trapping (Smart et a., 2015).

### *1.2.1 eDNA vs. capture-based sampling*

Fish assemblage monitoring is often conducted using field-based methods such as pulling seine nets and setting minnow traps (Lapointe et al., 2006). One study encompassing 400 km of the northern California and southern Oregon coastline compared the results of eDNA and seining for tidewater goby (*Eucyclogobius newberryi*) (Schmelzle, 2015). The authors found that the probability of fish detection using eDNA was almost double (0.74) compared to the detection with seining (0.39) (Schmelzle, 2015). This was a large-scale study of 29 locations at lagoon, slough, and estuary habitats with between 2 and

23 samples taken per location (Schmelzle, 2015).

Another study of coho salmon in the Smith River basin in California found that eDNA was an effective detection tool compared to visual count surveys (Shaffer, 2022). eDNA and visual counts had 93% agreement at the pool scale and 80% agreement at the reach scale (Shaffer, 2022). However, the author noted that eDNA was a poor predictor of coho salmon counts in the studied pools (Shaffer, 2022). Processing eDNA requires amplifying the small amount of collected DNA so there is enough data for sequencing (Kelly et al., 2019). The amplification does not affect all DNA equally, which results in an inaccurate representation of the abundance of fish species in the sample area (Kelly et al., 2019).

The eDNA sampling in the Elk River estuary is a restoration and monitoring effort on behalf of Caltrout. Caltrout is a non-profit organization with limited resources and personnel, and they did not have the required permits to conduct capture-based sampling in the study site. Under the Endangered Species Act, any form of "take" is prohibited for federally listed species, including survey, capture, and release (Fish and Wildlife Service, 2018). To conduct surveys for the tidewater goby, scientists must hold a Section 10(a)(1)(A) Recovery Permit granted by the U.S Fish and Wildlife Service (Fish and Wildlife Service, 2018). James Ray of the California Department of Fish and Wildlife was recruited to help with seining efforts as he has this permit. As mentioned in the article written by Evans et al., (2017), "fisheries are assessed by capture-based sampling, yet fisheries managers are unable to conduct extensive surveys due to budgetary constraints." eDNA sampling could combat this constraint and provide more accurate results for species detection, as eDNA requires less sampling effort and was 67% less expensive when compared to triple-pass electrofishing (Evans et al., 2017).

### 1.3 Fish assemblages in the Elk River estuary

The Elk River estuary in Humboldt County, California, provides habitat to over 30 known fish species (California Trout et al., 2023). Estuaries are extremely productive habitats because they have more abundant food sources and better temperatures for juvenile fish than stream-rearing habitats (Wallace et al., 2018). Additionally, they provide a transition area for smoltification that is neither the freshwater the fish were born in nor the saline ocean to which they are moving (Wallace et al., 2018). For these reasons, estuary-reared fish tend to be larger and more successful than their stream-reared counterparts (Wallace et al., 2018). Although stream-estuary ecotones can provide beneficial habitat, salmonid populations in the Elk River watershed have been declining (Humboldt Bay Watershed Advisory Committee, 2005). This is likely due to the impacts of legacy logging and ongoing agricultural practices resulting in sedimentation, decreased tidal influence, and loss of riparian vegetation (Humboldt Bay Watershed Advisory Committee, 2005). Even so, the Elk River estuary is home to many ecologically important species (California Trout et al., 2023).

### 1.4 Study objectives

The purpose of this study is to compare the efficacy of eDNA sampling in the Elk River estuary compared to capture-based sampling methods such as seining and minnow trapping. The non-profit organization California Trout (CalTrout) is interested in exploring the use of eDNA sampling to monitor fish assemblages before and after planned restoration efforts in the Elk River estuary system. For CalTrout, fisheries biodiversity would be a proxy for assessing ecosystem health as a result of restoration actions. Additionally, CalTrout is interested in testing if eDNA could be used to confirm the presence of state and federally listed threatened or endangered species within their defined critical habitat. These special status species are coho salmon (*Oncorhynchus kisutch*), Chinook salmon (*Oncorhynchus tshawytscha*), steelhead trout (*Oncorhynchus mykiss*), coastal cutthroat trout (*Oncorhynchus clarkii clarkii*), tidewater goby (*Eucyclogobius newberryi*), and longfin smelt (*Spirinchus thaleichthys*) (California Trout et al., 2023). Lastly, we will create an eDNA sampling schedule to monitor fish assemblages for CalTrout to implement in this area.

## 2. Pilot Study

### 2.1 Focal Species Background

#### *2.1.1 Coho salmon (Oncorhynchus kisutch)*

Coho juveniles prefer cool, slow-moving water that has cover, such as large woody structures, undercuts, or vegetation (California Trout et al., 2023). They can rear in fresh or brackish water (California Trout et al., 2023). Coho juveniles have been documented from March through July in the stream-estuary ecotone (California Trout et al., 2023). They utilize offstream habitat to overwinter in December and January, moving through the lower Elk River and Swain Slough to Martin Slough (Wallace & Allen, 2009). Salinity can play a significant role in the survival of coho smolts; they are typically found in water with a salinity of less than 5 ppt, but can survive up to 25 ppt (Wallace & Allen, 2009). However, salinities above 10 ppt result in decreased growth (Otto, 1971).

#### *2.1.2 Chinook salmon (Oncorhynchus tshawytscha)*

Juvenile Chinook salmon utilize the upper reach of the Elk River system from May to July and the lower reach from June to September (Wallace & Allen, 2009). They prefer to rear in the estuary and occupy salt marshes, sloughs, and the mainstem of the river for food and remain there until they are large enough to migrate to the ocean (NMFS, 2016). Juvenile Chinook have been found to move to smaller channels during mid- to late flood tides and out of the channel during ebb tides. Therefore, Chinook move between tidally influenced habitats (Hering et al., 2010). Chinook fry can withstand salinities less than 20 ppt, with osmoregulation increasing as they're exposed to brackish waters (Healey, 1991).

#### *2.1.3 Steelhead trout (Oncorhynchus mykiss)*

Steelhead require cool water with temperatures ranging from 44° F to 57° F and can face mortality in temperatures higher than that. According to a capstone study done by a senior at CSU Monterey Bay, a salinity of less than 10 ppt provides the most benefits for juveniles (Lumas, 2006). Juvenile steelhead occupy the upper reach of Elk River from May through October, February, and December and occupy the lower reach from March, April,

June, and July (Wallace & Allen, 2009). They grow quickly in the estuary and are assumed to have similar rearing requirements as other salmonids, requiring a slow moving, sheltered channel for protection (Bond et al., 2021). Spawning habitat is likely found in the Elk River, where they travel throughout the stream-estuary system to Swain and Martin Slough (California Trout et al., 2023).

### *2.1.4 Coastal cutthroat trout (Oncorhynchus clarkii)*

As with the salmonid species, Coastal cutthroat trout need cool water with cover and they prefer low gradient coastal streams and estuaries (California Department of Fish and Wildlife, 2018). Cutthroat exhibit various life history strategies depending on where certain populations decide to migrate. Some populations remain in stream systems, travel throughout various freshwater sections of riverways, or move between freshwater and saltwater (Moyle et al. 2015). Sea-run cutthroats will spawn upstream in freshwater systems and rely on estuaries for food source and habitat (Moyle et al. 2015). Since cutthroats are so diverse in aquatic environments, they can be found year round in the estuaries and lower reaches of stream systems (Moyle et al. 2015).

### *2.1.5 Tidewater goby (Eucyclogobius newberryi)*

Tidewater goby require slow moving, shallow brackish water to thrive. They can tolerate temperature and salinity fluctuations and live their whole lifecycle as part of the stream-estuary ecotone (Tetra Tech, Inc., 2000). Since Tidewater goby are so small, they are easily flushed downstream during high flood events (Lafferty et al., 1999). Juveniles and rearing adults prefer sand, mud, gravel, and silt that is generally associated with vegetated areas (Stillwater Sciences, 2006). Tidewater goby prefer water temperatures of 42° to 77° F with salinity below 15 ppt, although they can tolerate up to 51 ppt (Stillwater Sciences, 2006). Spawning occurs in spring and summer in disconnected lagoons, estuaries, and sloughs (California Trout et al., 2023). Previous studies comparing eDNA and seining show that eDNA sampling was almost twice as effective as seining in detecting Tidewater goby presence (Schmelzle, 2015). Tidewater goby are particularly susceptible to predation by other fish species and amphibians (Stillwater Sciences, 2006).

### *2.1.6 Longfin smelt (Spirinchus thaleichthys)*

Longfin smelt have a 2-year life cycle where they occur from nearshore waters to stream-estuary ecotones (Garwood 2017). Longfin smelt have been documented in all major tributaries in Humboldt Bay (Garwood 2017). Longfin smelt are thought to spawn in brackish-freshwater zones in upper portions of estuaries, with locations changing depending on freshwater inflow (Rosenfield 2010). While longfin smelt have not been captured in Elk River, it is assumed spawning occurs from late December through early March (California Trout et al., 2023). Longfin smelt larvae can tolerate salinities as low as 2 ppt and decline with salinities up to 15 ppt (Rosenfield, 2010). Overall, longfin smelt juveniles and pre-spawning adults can tolerate wide ranges of salinity but prefer 15-30 ppt (California Trout et al., 2023).

## 2.2 Study site description

The Elk River estuary (Figure 1) is located downstream of the Elk River Headwaters in

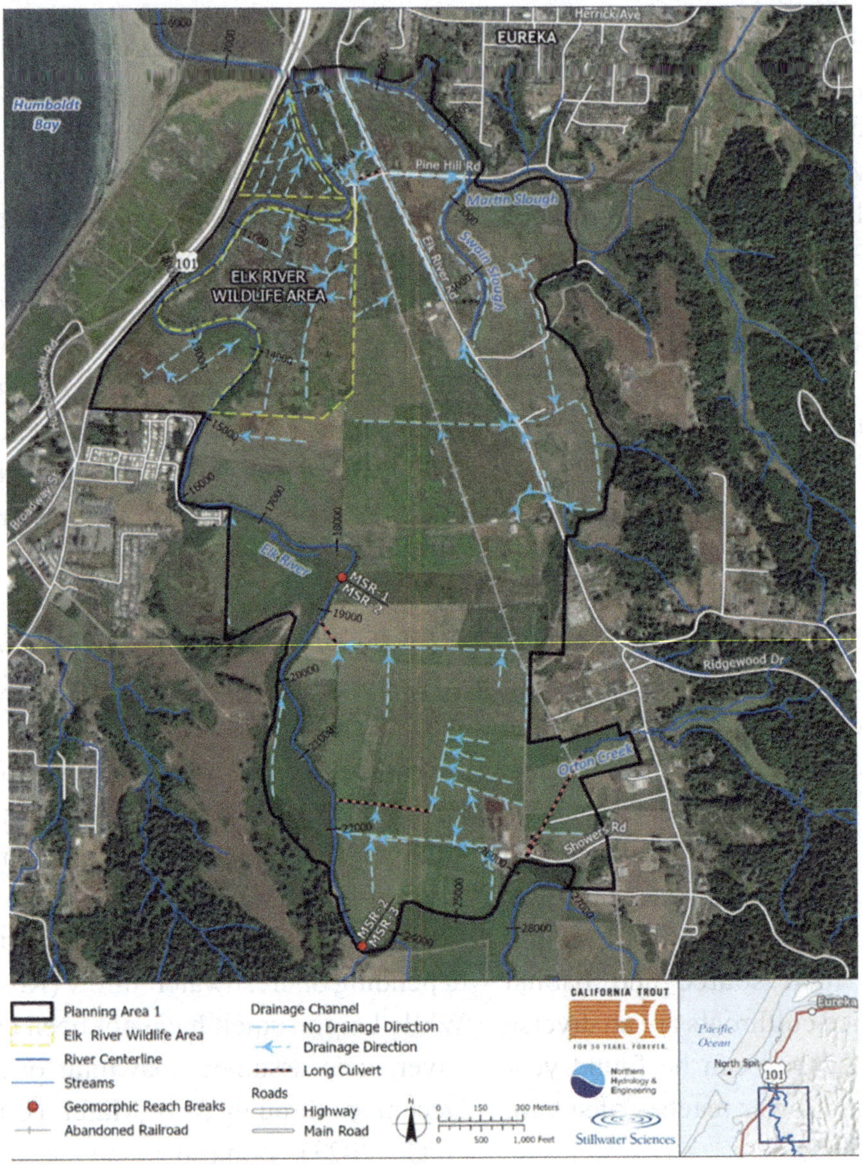


*Figure 1.* Map of the Elk River Estuary. California Trout's restoration project footprint is outlined in black. The pilot study was limited to mainstream reach 1, indicated on the map as downstream of the label "MSR-1". The Elk River Wildlife Area, an area of special interest, is outlined in yellow. (Source: California Trout).

Humboldt County, California, just south of Eureka, California. The estuary is also home to the Elk River Wildlife Area (ERWA). Our study area included 6 sampling sites where we conducted eDNA sampling and where CalTrout employees conducted seining and minnow trapping to use as a comparison (Figure 2).

## 2.3 Data collection

Seining, minnow trapping, and eDNA data were collected throughout the Elk River estuary at 6 total sites (Figure 2 and Table 1). Water samples for eDNA detection were collected on five dates (February 15, 2024, March 5, 2024, March 11,

*Figure 2. Map of sampling sites for eDNA, seining and minnow trapping for this study. Sites are labeled with the associated stream and CalTrout numerical delineation. Made by Christy Wheatley using ArcGIS Pro version 3.2.2.*

2024, and April 8, 2024). The number of sites visited varied by day, with two samples taken at each site. A total of 32 eDNA samples were taken throughout the course of the study. Water quality measurements, including salinity, temperature, and dissolved oxygen were also taken during each sampling occasion.

### *2.3.1 eDNA data collection instructions*

Prior to collecting water samples, label each Whirl-Pak bag with the date, site code, and sample number and record the information on a datasheet or notebook. For this project, we used single use, Nasco Whirl-Pak bags. Put on sterile disposable gloves (latex or nitrile) to avoid contaminating

*Table 1.* Sampling sites and what methods were conducted in each. "X" represents the method conducted at each site. Sites are labeled by the waterbody and CalTrout's numbering standard.

| Site | eDNA sampling | Seine netting | Minnow trapping |
|---|---|---|---|
| Elk River 1.5 | X | X | |
| Elk River 1 | | X | X |
| Elk River 3 | X | | X |
| Swain Slough 2 | X | | |
| Swain Slough 1.5 | X | X | |
| Side Channel 2 | X | | |

the samples. If the gloves become contaminated either change gloves or rinse with RNase-away. Open the labeled Whirl-Pak bag by removing the protective plastic strip at the top of the bag. Use the plastic pull tabs located on either side of the opening to fully open the bag while avoiding contact with the sterilized interior. To collect the water sample, grab the bag by pinching along one side of the opening. Dip the opposite side of the Whirl-Pak bag into the water and fill the bag approximately ¾ full, or about 750mL (Figure 3). Ensure you are not downstream from possible contaminants (swimmers, fishermen, boats, etc.). Avoid sampling near back eddies and suspended sediment to reduce the amount of sediment in the samples.

To close the Whirl-Pak bag, roll the opening of the bag four to five times and secure it with the integrated tabs. Repeat these steps for each water sample. Once all the samples have been collected, begin filtering the samples in the field or transport the samples to a designated location for filtration. If transporting the water samples, be sure to place the bags in a cooler with ice if the samples will

*Figure 3. Collecting an eDNA sample. It is easiest to collect the water without contaminating the bag by pinching the bag open as shown.*

not be filtered within 30 minutes of collection. Ensure the bags are in an upright position to avoid leakage and are not exposed to UV. For the most accurate results the samples should be filtered within 8 hours of collection. Prior to filtering, label sample tubes (cap and side of tube) with the date, site code, and sample number (Figure 4). There should be a corresponding tube and Whirl-Pak bag for each sample. This pilot study used 2 ml microcentrifuge tubes (Eppendorf catalog number: 022431048) containing 360 µL of cell lysis buffer (QIAGEN buffer ATL, catalog number: 939011).

Begin setting up the filtration manifold and vacuum pump in a clean, level location away from hazards to avoid contamination. Connect the manual vacuum pump to the adapter on the filtration manifold (created by Jason Shaffer) with silicone tubing (Figure 5). Apply a new set of disposable sterile gloves, then fit the base of the sterilized filter funnel(s) into the rubber stopper hole on the filtration manifold (Figure 6). This pilot study used Thermo Scientific™ Nalgene™ Single-Use Analytical Filter Funnels. Ensure it is snug and leakproof but do not touch the inside of the funnel. Each funnel should only filter water from a single sample to avoid contamination and incorrect results.

Open the manifold adapters for each filter cup that will contain the water samples (Figure 7).

*Figure 5. Connecting the vacuum pump to the manifold.*

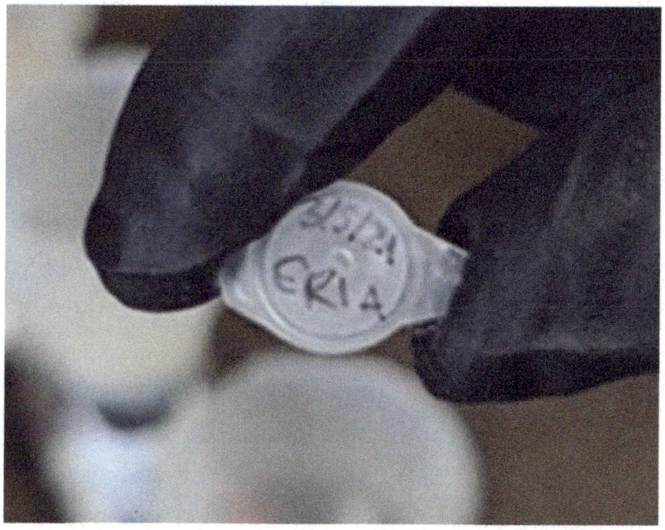

*Figure 4. Example tube label with data, sampling location, and sample number.*

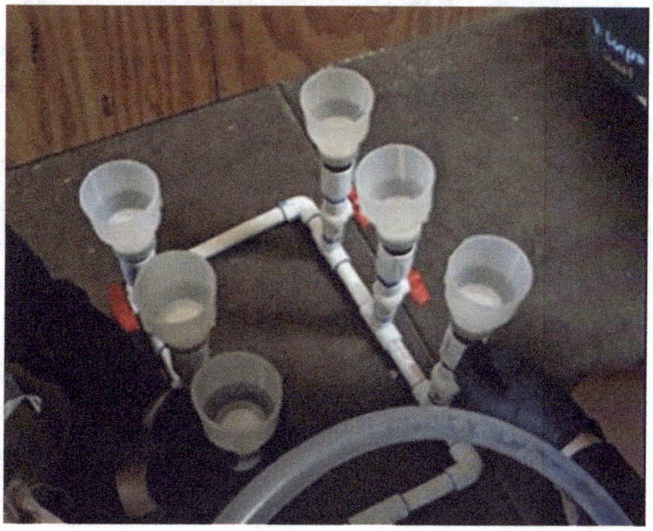

*Figure 6. Filtration manifold with filter cups attached.*

Unroll the Whirl-Pak sample bag and fill each cup to the 250 mL line, making sure nothing touches the inside of the filter except the water sample (Figure 8).

Add a field blank of distilled water into one filter cup as a contamination control and filter simultaneously with the collected samples. It is only necessary to use 250-500mL of water for the field blank. Begin manually pumping the vacuum. The speed at which the water is filtered makes no difference, so pumping quickly or vigorously is unnecessary and will lead to fatigue. As the samples are filtered out of the funnel, slowly pour the remainder of the water sample into the filter funnel until it is at the 250mL line. To track the amount of water filtered, record the number of cups filled for each sample, then multiply by 250ml.

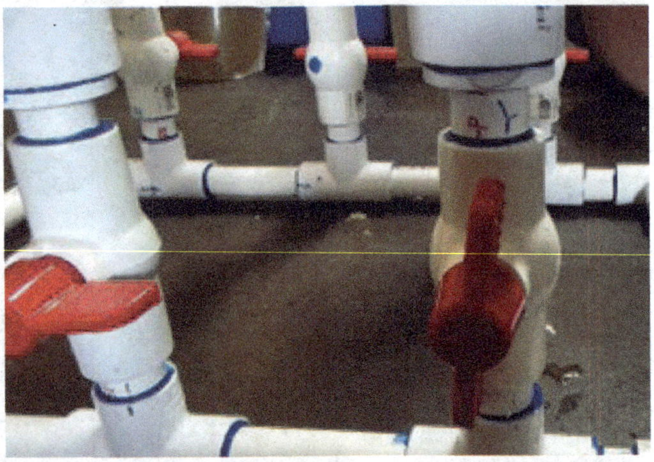

*Figure 7. Vertical adapters are open (right), and horizontal adapters are closed (left).*

*Figure 8. Fill the filter cups with the sample, taking care to avoid touching the cup.*

Once all the water samples have been fully filtered, put on a new set of sterile gloves. Remove the top of the filter funnel to expose the filter membrane. This pilot study used 0.45-micron cellulose nitrate filters held in the base of the filter funnels. Sterilize a pair of forceps by dipping the forceps into a 10% bleach solution, then water, three times. Always end the third dunk in water to avoid bleaching and contaminating the filtered samples. Using the sterilized forceps, carefully fold the filter membrane in half three times, making sure the face of the membrane containing the DNA sample is folded in on itself (Figure 9). Place the folded filter into the labeled microcentrifuge tube with preservation media, such as cell lysis buffer. Once all the samples have been filtered and the membrane is stored in a microcentrifuge tube, release the pressure in the pump using the handle at the top and disconnect the pump from the manifold. For a video tutorial of this eDNA sampling procedure, see this link: https://youtu.be/f2a3dugLrYA.

After filtering, the samples were sent to Dr. Andrew Kinziger's laboratory at Cal Poly Humboldt for DNA extraction and metabarcoding using MiFish primers. The lab also ran negative and positive PCR controls to amplify the eDNA for detection.

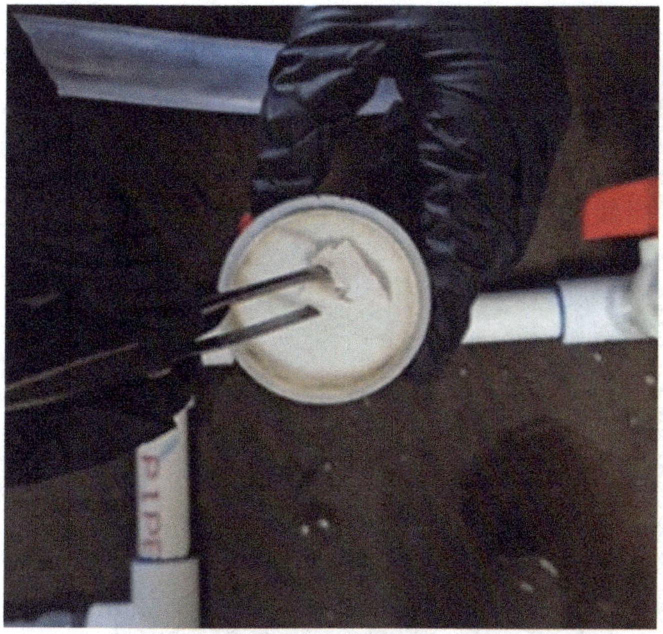

*Figure 9. The filter should look like a triangle after the folding process.*

If the water is not pumping through the filters, first, check that all manifold adapters are in the proper position. Those with cups should be open, and those without cups should be closed. If you have already pumped a lot of water, check to ensure the pump tank isn't full. If it is, release the pressure, disconnect it from the manifold, and pour out the filtered water before starting again. If you are filtering water with a lot of sediment, the filter may become clogged before all the water has filtered through. If the filter paper rips as you're folding it, make sure that you put all the pieces in the microcentrifuge tube.

#### *2.3.2 Capture-based sampling methods*

Seining and minnow trapping were conducted on three dates to provide comparison data for eDNA results (March 8, 2024, March 11, 2024, and April 8, 2024). Two of these dates overlapped with our eDNA sampling. The locations of seining and minnow trapping did not overlap entirely with eDNA sampling sites due to the difficult logistics of seining in some parts of the estuary. However, our eDNA samples are likely to pick up a wide range of data due to the influence of the tide throughout the estuary (Baetscher et al., 2024). Matthew Methany of Caltrout and James Ray of California Department of Fish and Wildlife pulled seines at four sample sites. They used 8-foot long by 4-foot deep pole seines with a 3 mm mesh and a 40-foot long by 6-foot deep seine with a 3 mm mesh deployed via kayak.

### 2.4 Proposed data analysis

Data analysis will be conducted to compare eDNA metabarcoding and capture-based sampling methods once the eDNA results have been received in May. Linear mixed effects modeling will be used to compare the abundance of species detected with each method using lme4 in R studio (He et al., 2022). Bayesian modeling will be conducted with the bmrs package in R studio to see if any of the water quality measurements have a statistically significant impact on the eDNA results (C. Dixon, pers. comm, 2024).

## 3. Monitoring Plan

### 3.1 Proposed sampling schedule

The following document contains our recommendations for eDNA sampling in the Elk River estuary system. Due to delays in lab processing,

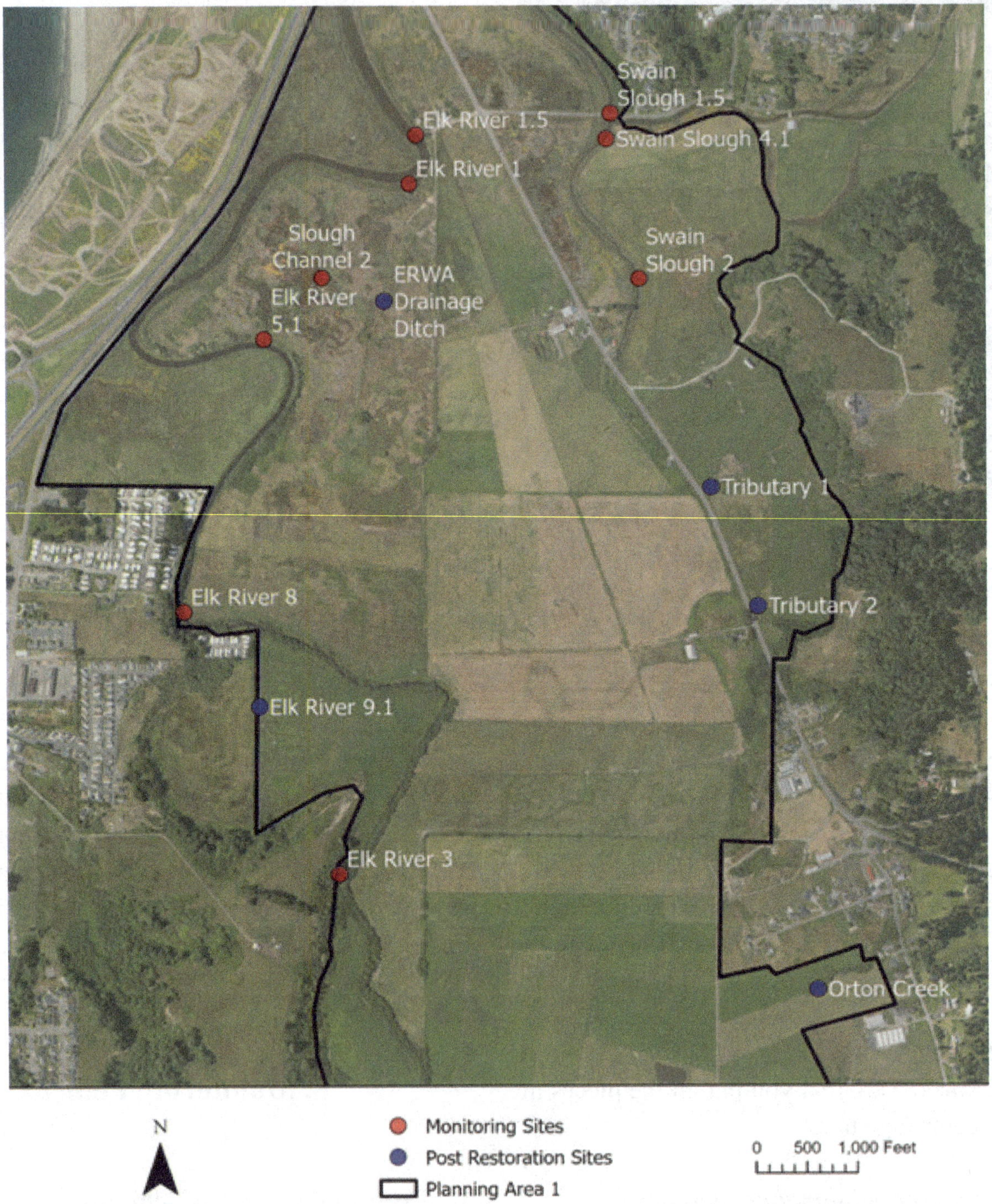


*Figure 10. Map of proposed monitoring locations in the Elk River estuary. Initial sites are shown in red, while post-restoration sites are marked with blue. Made by Christy Wheatley using ArcGIS Pro version 3.2.2.*

*Table 2.* Recommended sampling sites per calendar month with the focal species that are expected in the Elk River estuary.

| Month | Sites to be Visited | Expected Species |
|---|---|---|
| January | Elk River 1.5 tide gate | coho salmon |
| | Elk River 1 | coho salmon |
| | Elk River 8 | coho salmon |
| | Swain Slough 1.5 | tidewater goby<br>longfin Smelt |
| | ERWA slough channel | longfin smelt |
| February | Swain Slough 1.5 | tidewater goby<br>longfin smelt |
| | ERWA slough channel (SC 2) | longfin smelt |
| March | Elk River 1.5 tide gate | coho salmon<br>steelhead trout |
| | Elk River 8 | coho salmon<br>steelhead trout |
| | Swain Slough 1.5 | tidewater goby<br>longfin smelt |
| | ERWA slough channel (SC 2) | longfin smelt |
| April | Elk River 1.5 | coho salmon<br>steelhead trout |
| | Elk River 8 | coho salmon<br>steelhead trout |
| | Swain Slough 1.5 | coho salmon<br>steelhead trout<br>tidewater goby<br>longfin smelt |
| | Swain Slough 4.1 Alcove | tidewater goby |
| May | Elk River 5.1 | coho salmon<br>coastal cutthroat trout<br>tidewater goby |
| | Swain Slough 4.1 Alcove | tidewater goby |
| June | Elk River 5.1 | coho salmon<br>Chinook salmon<br>steelhead trout<br>coastal cutthroat trout |
| | Swain Slough 4.1 Alcove | tidewater goby |

*Table 2 continued on next page*

| Month | Sites to be Visited | Expected Species |
|---|---|---|
| July | Elk River 5.1 | coho salmon<br>Chinook salmon<br>steelhead trout<br>coastal cutthroat trout |
| | Swain Slough 4.1 Alcove | tidewater goby |
| August | Elk River 5.1 | Chinook salmon<br>tidewater goby |
| | Swain Slough 4.1 Alcove | tidewater goby |
| September | Elk River 1.5 | Chinook salmon |
| | Swain Slough 1.5 | tidewater goby |
| October | Swain Slough 1.5 | tidewater goby |
| November | Swain Slough 1.5 | tidewater goby |
| December | Elk River 1.5 | coho salmon |
| | Swain Slough 1.5 | tidewater goby<br>longfin smelt |

the eDNA results were not available by the end of this project in May 2024. Thus, sampling sites and timing (Table 2) were determined using a literature review and known locations of focal species within the estuary from previous seining. Unless otherwise stated, eDNA sampling is likely to be most effective at high tides because that is when high-quality rearing habitats will be accessible to fish (California Trout et al., 2023).

### 3.2 Post-restoration monitoring

Some areas are likely to provide habitat for fish species, but are currently disconnected from the main channel and are inaccessible. These areas are candidates for post-restoration monitoring. Table 3 includes the sites and the species likely to be found therein. The site locations include ER 9.1, which is noted as a potential high-quality habitat. The Elk River Wildlife Area (ERWA) has a series of ponds and historic sloughs that could provide habitat but are currently low-quality and generally cut off from tidal influence. The tributaries T1 and T2, which flow into Swain Slough, could provide freshwater habitat. Orton Creek has the potential to provide spawning habitat for coastal cutthroat trout.

### 3.3 Sampling Budget

Samples are processed at a cost of $140 per sample. Each site will require the collection of 2 samples. There are 29 proposed sites, which comes to a total of 58 samples to be collected throughout one year of monitoring. This comes to a total cost of $7,840. This work can be conducted by fisheries technicians, whereas biologists who have permits to conduct capture-based methods are typically

*Table 3.* Prospective locations for post-restoration eDNA sampling along with the expected focal species in the Elk River estuary. See Figure 10 for a map of site locations.

| Site to be Visited | Expected Species |
|---|---|
| ER 9.1 | coho salmon<br>Chinook salmon<br>steelhead trout<br>coastal cutthroat trout<br>tidewater goby |
| ERWA drainage ditches | coho salmon<br>Chinook salmon<br>steelhead trout<br>coastal cutthroat trout<br>tidewater goby<br>longfin smelt |
| T1 & T2 | coho salmon<br>Chinook salmon<br>steelhead trout<br>coastal cutthroat trout |
| Orton Creek | coastal cutthroat trout |

paid two to three times more than technicians (State of California, 2024).

## 3.4 Considerations for eDNA sampling in the Elk River

A habitat limitation in the Elk River for fish is a "lack of in-channel cover and velocity refugia during higher flows," which directly impacts fish as they require low flows and off-channel ponds to conserve energy (California Trout et al., 2023). If the Elk River continues to lose habitat, it will also lose biodiversity of these focal fish species. Additionally, it's important to consider where fish are likely to inhabit when choosing monitoring sites (J. Ray, pers. comm, 2024). Accessibility can be an issue with sampling the Elk River as most sites are private property and require landowner permission to access. Even with permission, the banks of the sloughs are extremely muddy and there is a high risk of slipping and falling into the channel.

The historical logging and agriculture practices of the Elk River watershed have contributed to consistently turbid water in the Elk River system (Klein et al., 2012). When filtering the collected samples, excessive sediment "plugs up" the filter membrane and makes it difficult to filter all of the water collected. In this case, it may be helpful to have extra filter membranes available in the field to ensure enough of the water sample has been filtered. There are conflicting results on how sediment impacts DNA samples; some sources found that "eDNA accumulated and decayed more slowly in sediment than

water" (Nevers, et al., 2020), while others found that detection sensitivity was negatively affected by sediment (Holmes et al., 2024).

## 4. Recommendations

Based on our pilot study and monitoring plan, we recommend that CalTrout continue using and comparing eDNA sampling against traditional methods. While our pilot study found that eDNA sampling was able to be conducted on multiple sites in the Elk River Estuary and was less labor intensive, it's important to continuously evaluate the effectiveness for eDNA sampling against traditional methods. Additional eDNA sampling should be conducted before, during, and after the restoration to accurately assess its effectiveness. We also suggest monitoring a reference site where restoration is not planned to compare the results of the restoration. Two suggested areas are Swain Slough 4.1 and Swain Slough 2. Our monitoring plan serves as a starting point, but monitoring should be adaptable in order to be effective; sampling should take place where fish are expected to be and that could change as the restoration progresses. By incorporating these recommendations into ongoing monitoring efforts, CalTrout can increase the effectiveness of eDNA sampling for monitoring fish presence and develop valuable data for estuary restoration.

## 5. Acknowledgements

The authors thank California Trout for the funding and guidance provided on this project, especially to Katy Gurin and Matthew Methany. We are also grateful to the assistance provided by Dr. Alison O'Dowd and Dr. Daniel Lipe. Lastly, we thank James Ray of the California Department of Fish and Wildlife for making our seining efforts possible.

## 6. Literature Cited

Baetscher, D. S., Pochardt, M. R., Barry, P. D., & Larson, W. A. (2024). Tide impacts the dispersion of eDNA from nearshore net pens in a dynamic high-latitude marine environment. *Environmental DNA*, *6*(2), e533. https://doi.org/10.1002/edn3.533

Bash, J. S., & Ryan, C. M. (2002). Stream Restoration and Enhancement Projects: Is Anyone Monitoring? *Environmental Management*, *29*(6), 877–885. https://doi.org/10.1007/s00267-001-0066-3

Beng, K. C., & Corlett, R. T. (2020). Applications of environmental DNA (eDNA) in ecology and conservation: opportunities, challenges and prospects. *Biodiversity and Conservation*, *29*(7), 2089–2121. https://doi.org/10.1007/s10531-020-01980-0

Bennett, S., Pess, G., Bouwes, N., Roni, P., Bilby, R. E., Gallagher, S., Ruzycki, J., Buehrens, T., Krueger, K., Ehinger, W., Anderson, J., Jordan, C., Bowersox, B., & Greene, C. (2016). Progress and Challenges of Testing the Effectiveness of Stream Restoration in the Pacific Northwest Using Intensively Monitored Watersheds. *Fisheries*, *41*(2), 92–103. https://doi.org/10.1080/03632415.2015.1127805

Bond, M. H., Hayes, S. A., Hanson, C. V., & MacFarlane, R. B. (2008). Marine survival of steelhead (Oncorhynchus mykiss) enhanced by a seasonally closed estuary. *Canadian Journal of Fisheries and Aquatic Sciences*, *65*(10), 2242–2252. https://doi.org/10.1139/F08-131

California Department of Fish and Wildlife. (2018). Coastal Cutthroat Trout. Retrieved from https://www.calfish.org/FisheriesManagement/SpeciesPages/CoastalCutthroatTrout.aspx

California Water Boards. (2012). Final California 2012 Integrated Report Supporting Information. Waterboards.ca.gov. https://www.waterboards.ca.gov/water_issues/programs/tmdl/2012state_ir_reports/03308.shtml#32333

California Trout, Northern Hydrology and Engineering, & Stillwater Sciences. (2023). Elk River Planning Area 1: 10% Design Report.

California Water Boards. (2012). Final 2012 Integrated Report (CWA Section 303(d) List / 305(b) Report). https://www.waterboards.ca.gov/water_issues/programs/tmdl/2012state_ir_reports/category5_report.shtml

DiBattista, J. D., Fowler, A. M., Riley, I. J., Reader, S., Hay, A., Parkinson, K., & Hobbs, J.-P. A. (2022). The use of environmental DNA to monitor impacted coastal estuaries. *Marine Pollution Bulletin*, *181*, 113860. https://doi.org/10.1016/j.marpolbul.2022.113860

Evans, N. T., Shirey, P. D., Wieringa, J. G., Mahon, A. R., & Lamberti, G. A. (2017). Comparative Cost and Effort of Fish Distribution Detection via Environmental DNA Analysis and Electrofishing. *Fisheries*, *42*(2), 90–99. https://doi.org/10.1080/03632415.2017.1276329

Finegold, C. (2009). *The importance of fisheries and aquaculture to development*. The Royal Swedish Academy of Agriculture and Forestry. https://digitalarchive.worldfishcenter.org/handle/20.500.12348/1445

Garwood, R. S. (2017). Historic and contemporary distribution of Longfin Smelt (Spirinchus thaleichthys) along the California coast. *California Fish and Game*, *103*(3), 96–117. https://doi.org/https://nrm.dfg.ca.gov/FileHandler.ashx?DocumentID=152480&inline

He, X., Stanley, R. R. E., Rubidge, E. M., Jeffery, N. W., Hamilton, L. C., Westfall, K. M., Gilmore, S. R., Roux, L.-M. D., Gale, K. S. P., Heaslip, S. G., Steeves, R., & Abbott, C. L. (2022). Fish community surveys in eelgrass beds using both eDNA metabarcoding and seining: implications for biodiversity monitoring in the coastal zone. *Canadian Journal of Fisheries and Aquatic Sciences*, *79*(8), 1335–1346. https://doi.org/10.1139/cjfas-2021-0215

Hering, D. K., Bottom, D. L., Prentice, E. F., Jones, K. K., & Fleming, I. A. (2010). Tidal movements and residency of subyearling Chinook salmon (Oncorhynchus tshawytscha) in an Oregon salt marsh channel. *Canadian Journal of Fisheries and Aquatic Sciences*, *67*(3), 524–533. https://doi.org/10.1139/F10-003

Hiddink, J. G., MacKenzie, B. R., Rijnsdorp, A., Dulvy, N. K., Nielsen, E. E., Bekkevold, D., Heino, M., Lorance, P., & Ojaveer, H. (2008). Importance of fish biodiversity for the management of fisheries and ecosystems. *Fisheries Research*, *90*(1), 6–8. https://doi.org/10.1016/j.fishres.2007.11.025

Holmes, A., Baerwald, M., Rodzen, J., Schreier, B., Mahardja, B., & Finger, A. (2024). Evaluating environmental DNA detection of a rare fish in turbid water using field and experimental approaches. *PeerJ*. https://escholarship.org/uc/item/9929m0kv

Humboldt Bay Watershed Advisory Committee. (2005). Humboldt Bay Watershed Salmon and Steelhead Conservation Plan. http://www.nrsrcaa.org/nrs/projcurr/pdfs/EnhancePlanComplete.pdf

Joseph, C., Faiq, M. E., Li, Z., & Chen, G. (2022).

Persistence and degradation dynamics of eDNA affected by environmental factors in aquatic ecosystems. *Hydrobiologia*, *849*(19), 4119–4133. https://doi.org/10.1007/s10750-022-04959-w

Kelly, R. P., Shelton, A. O., & Gallego, R. (2019). Understanding PCR Processes to Draw Meaningful Conclusions from Environmental DNA Studies. *Scientific Reports*, *9*(1), 12133. https://doi.org/10.1038/s41598-019-48546-x

Klamt, R. (n.d.). *Elk River Watershed.* North Coast Regional Water Quality Control Board. Retrieved March 23, 2024, from https://www.waterboards.ca.gov/northcoast/water_issues/programs/watershed_info/eureka_plain/elk/

Klein, R. D., Lewis, J., & Buffleben, M. S. (2012). Logging and turbidity in the coastal watersheds of northern California. *Geomorphology*, *139–140*, 136–144. https://doi.org/10.1016/j.geomorph.2011.10.011

Lafferty, K. D., Swift, C. C., & Ambrose, R. F. (1999). Post flood persistence and recolonization of endangered tidewater goby populations. *North American Journal of Fisheries Management*, *19*, 618–622. https://pubs.usgs.gov/publication/1008148

Lapointe, N. W. R., Corkum, L. D., & Mandrak, N. E. (2006). A Comparison of Methods for Sampling Fish Diversity in Shallow Offshore Waters of Large Rivers. *North American Journal of Fisheries Management*, *26*(3), 503–513. https://doi.org/10.1577/M05-091.1

Lumas, E. (2006). Water quality monitoring for suitable juvenile Steelhead trout (Oncorhynchus mykiss) habitat in the Carmel River Lagoon during minimal fresh water inflow : a capstone project. *Capstone Projects and Master's Theses*. https://digitalcommons.csumb.edu/caps_thes/51

"Minimum Qualifications for Obtaining a Section 10(a)(1)(A) Recovery Permit for Conducting Surveys for the Tidewater Goby (*Eucylogobius newberryi*), (2018, June 3). U.S Fish and Wildlife Service. https://www.fws.gov/sites/default/files/documents/minimum-requirements-for-tidewater-goby.pdf

Moyle, P.B., Leidy, R.A. (1992). Loss of Biodiversity in Aquatic Ecosystems: Evidence from Fish Faunas. In: Fiedler, P.L., Jain, S.K. (eds) Conservation Biology. Springer, Boston, MA. https://doi.org/10.1007/978-1-4684-6426-9_6

Moyle, P. B., Quiñones, R. M., Katz, J. V., & Weaver, J. (2015). *Fish Species of Special Concern in California*. California Department of Fish and Wildlife. https://nrm.dfg.ca.gov/FileHandler.ashx?DocumentID=104268

Nagarajan, R. P., Bedwell, M., Holmes, A. E., Sanches, T., Acuña, S., Baerwald, M., Barnes, M. A., Blankenship, S., Connon, R. E., Deiner, K., Gille, D., Goldberg, C. S., Hunter, M. E., Jerde, C. L., Luikart, G., Meyer, R. S., Watts, A., & Schreier, A. (2022). Environmental DNA Methods for Ecological Monitoring and Biodiversity Assessment in Estuaries. *Estuaries and Coasts*, *45*(7), 2254–2273. https://doi.org/10.1007/s12237-022-01080-y

National Marine Fisheries Service. (2016). *Coastal Multispecies Final Recovery Plan: California Coastal Chinook Salmon ESU, Northern California Steelhead DPS and Central California Coast Steelhead DPS.* West Coast Region, Santa Rosa, California.

Nevers, M. B., Przybyla-Kelly, K., Shively, D., Morris, C. C., Dickey, J., & Byappanahalli, M. N. (2020). Influence of sediment and stream transport on detecting a source of environmental DNA. PloS one, 15(12), e0244086. https://doi.org/10.1371/journal.pone.0244086

Otto, R. G. (1971). Effects of Salinity on the Survival and Growth of Pre-Smolt Coho Salmon

(Oncorhynchus kisutch). *Journal of the Fisheries Research Board of Canada*, *28*(3), 343–349. https://doi.org/10.1139/f71-046

Rees, H.C., Maddison, B.C., Middleditch, D.J., Patmore, J.R.M. and Gough, K.C. (2014), REVIEW: The detection of aquatic animal species using environmental DNA – a review of eDNA as a survey tool in ecology. J Appl Ecol, 51: 1450-1459. https://doi.org/10.1111/1365-2664.12306

Roni, P., Hanson, K., & Beechie, T. (2008). Global Review of the Physical and Biological Effectiveness of Stream Habitat Rehabilitation Techniques. *North American Journal of Fisheries Management*, *28*(3), 856–890. https://doi.org/10.1577/M06-169.1

Rosenfield, J. A. (2010). *Life History Conceptual Model and SubModels for Longfin Smelt, San Francisco Estuary Population* [Supporting Text]. Delta Regional Ecosystem Restoration Implementation Plan. https://nrm.dfg.ca.gov/FileHandler.ashx?DocumentID=28421

Ruppert, K. M., Kline, R. J., & Rahman, M. S. (2019). Past, present, and future perspectives of environmental DNA (eDNA) metabarcoding: A systematic review in methods, monitoring, and applications of global eDNA. *Global Ecology and Conservation*, *17*, e00547. https://doi.org/10.1016/j.gecco.2019.e00547

Schmelzle, M. (2015). *Using occupancy modeling to compare environmental DNA to traditional field methods for regional-scale monitoring of an endangered aquatic species* [Masters Thesis, Humboldt State University]. https://scholarworks.calstate.edu/downloads/736666889

Shaffer, J. (2022). *Comparison of environmental DNA and underwater visual count surveys for detecting juvenile coho salmon (Oncorhynchus kisutch) in rivers* [Masters Thesis, California State Polytechnic University, Humboldt]. https://digitalcommons.humboldt.edu/cgi/viewcontent.cgi?article=1656&context=etd

Smart, A. S., Tingley, R., Weeks, A. R., van Rooyen, A. R., & McCarthy, M. A. (2015). Environmental DNA sampling is more sensitive than a traditional survey technique for detecting an aquatic invader. Ecological Applications, 25(7), 1944–1952. https://doi.org/10.1890/14-1751.1

Stillwater Sciences. (2006). *Guidelines to Evaluate, Modify and Develop Estuarine Restoration Projects for Tidewater Goby Habitat* [Final Report]. https://s3-us-west-2.amazonaws.com/ucldc-nuxeo-ref-media/f8522018-eadc-4564-877f-789cc89438c3

State of California. (2024). *Civil Service Pay Scale*. California Department of Human Resources. https://calhr.ca.gov/Pay%20Scales%20Library/PS_Sec_15.pdf

Tetra Tech, Inc. (2000). *Intensive habitat survey for Lake Earl and Lake Talawa, Del Norte County, California*. [Final Report]. US Army Corps of Engineers.

Thomas, A. C., Howard, J., Nguyen, P. L., Seimon, T. A., & Goldberg, C. S. (2018). eDNA Sampler: A fully integrated environmental DNA sampling system. *Methods in Ecology and Evolution*, *9*(6), 1379–1385. https://doi.org/10.1111/2041-210X.12994

Wallace, M., Ojerholm, E., Scheiff, A., & Allen, S. (2018). *Juvenile Salmonid Use And Restoration Assessment Of The Tidal Portions Of Selected Tributaries To Humboldt Bay, California* [Fisheries Restoration Grant Program Final Report]. California Department of Fish and Wildlife. https://nrm.dfg.ca.gov/FileHandler.ashx?DocumentID=17970Wh

IdeaFest
Journal

# Finding Joy and Community in Climate Action

**Deanna Kellom**

## Abstract

The goal of this paper is to discuss the ways that the conversation around climate action can be altered to center finding joy and community. Often when people hear about the state of the climate there is a sense of dread that people have reported feeling. Most people feel as though the climate has reached the point of no return and that there is nothing that any one person can do to really make a difference. This paper will talk about some ways that have worked for people in getting them to do climate action while also enjoying the work that needs to be done. Some topics that will be discussed are environmental communication as well as climate action in specific communities. There will also be an analysis on how the conversation of climate change has been handled in the past and why it makes people feel as though there is no hope. By using examples of communities that have found a way to participate in climate action, the goal will be that everyone can feel like they can take part in caring for the climate while also caring for themselves.

Can there be joy in climate action? When people hear about climate change and all the drastic events that come with it, it is easy to feel like they are doomed and that there is nothing anyone can do to stop it. The anxiety that happens around climate change is one of the reasons the discourse needs to shift to how joy can be related to climate action. Instead of feeling every consequence to climate change at once, there is a way that people can do their part while also being happy. In the article, "Happy Climate Is On a Mission to Show What We Gain From Climate Action", they talk about what Lise Van Susteren thinks about climate action. The article states that, "'One of the best ways to decrease anxiety is to engage in empowering actions — that means joining groups that are doing stuff,' says Van Susteren. This observation is borne out in a recent study of university students, which found that taking collective actions can serve as a buffer for climate depression," (Parshley 2022). This illustrates how there is a need to move away from language that makes us feel anxious for our planet. It also shows the importance of collective actions and how community is something that people can lean on to stay resilient while participating in climate action. This paper will discuss the ways that society can reframe the conversation around climate change to include the things that we can do that make people happy as well as make a difference.

Ayana Elizabeth Johnson who is a marine biologist, writer, and climate policy expert, put on a Ted Talk called "How to Find Joy in Climate Action". Johnson mentions that, "the climate challenge is gargantuan, thanks largely to fossil fuel executives and the PR firms and politicians doing their bidding" (Johnson 0:26). In her Ted Talk, she talks about climate action in a way that puts a focus on how everyone can use their abilities to fight climate change. This example is interesting because she makes climate action seem attainable and if people can come together, they can make a big difference. She also brings to light some things that people should know about. In her Ted Talk she says that the term "carbon footprint" was "popularized by fossil fuel corporations in an attempt to put the blame on us as individuals" (Johnson, 1:25). This is enlightening because people hear about how they can minimize their carbon footprint, but not how big companies are doing their part to stop their fossil fuel emissions. She then puts the emphasis on using the skills that people already have to create change. She does this by having a venn diagram that includes 3 questions: What are you good at, what is the work that needs doing, and what brings you joy or satisfaction? The reason for this approach is because it will yield more progress on climate solutions than if people were only focused on their own impact.

This example is interesting because she talks about climate action in a new way by saying that people can find joy in climate action. This is an approach that not many people are using and when you think of climate change, most people aren't really thinking about how you can find joy in it. She emphasizes the importance of having doers, not just people that will talk about the fact that climate action needs to be taken seriously. This is important to talk about the need for doers because only talking about the crisis doesn't equal action. Asking people what they are good at and what they like doing can help in the long run for many reasons. It is important that people don't lose their momentum when participating in climate action by doing something that they do not enjoy. It is much easier to continue doing something when enjoying doing it, so asking that question can help people to be more resilient with the work that they are doing. It is also important to ask what people are good at because we want people to be confident in their abilities in order to see change.

Additionally, people need to stop using scare tactics to get people to act. It usually creates a sense that the world is in such a bad place that people feel that even if they do something, it still won't be enough. A lot of the new generation that is just now getting involved in the conversation about climate change, views it as a gargantuan issue that can not be solved. They feel as though this issue was forced upon them by previous generations and big companies and there is not one solution to stopping the issue at hand. That is why reframing the conversation is necessary because it is often about scaring people into caring, when it can be about doing something that makes you happy and also makes a difference. Of course, people should stay informed on news about the climate and climate scientists should remain truthful to the impacts that it will have on humanity. The role of being the one to share scientific knowledge is large and it is important that it is done carefully and thoughtfully, which is why scientists can take ideas from rhetoricians to develop new methods of relaying climate information and persuading people to act. Another thing that scientists can do is move the conversation from talking about the problems to talking about the solutions. In an interview with Chris Jordan, he states that "everybody is sick of hearing about the problems, so the focus now is only on 'solutions', not complex global solutions involving the takedown of capitalism and renewed spiritual connection with life, but 'solutions' limited to individual behavioral gestures that everyone knows don't address the scale of the problems by even an order of magnitude," (Chris, 116). I think that this is a really important point to bring up, that even if everyone was to make those changes in their life, there is still going to be a climate crisis. There are a lot of changes that need to be made on a larger scale but it is hard to know about what needs to be done when you only have an outside perspective.

While talking about changing the conversation around climate action, another important aspect that can be focused on is community engagement. To make people feel as though they are not alone in fixing the climate crisis, they can get involved in community initiatives to not only help out, but to connect with others. There are many examples of how being active in a community is better than any one person making an effort. One example of this is from a review done by Maria João Salvador Costa, Alexandra Leitão, Rosa Silva, Vanessa Monteiro, and Pedro Melo titled, "Climate Change Prevention through Community Actions and Empowerment: A Scoping Review", in which they talk about a variety of different communities and their climate actions. They also discuss the importance of utilizing both a top-down approach and a bottom-up approach to address community needs and environmental intervention. They state that, "Participation processes are required to involve several stakeholders in assessing community resilience and have significant benefits as they can effectively raise awareness and broaden the understanding of risks, promoting local participation at the same time [6]." (Salvador Costa, et al. 2022). This review shows the importance of community involvement and how climate action has worked for certain communities. This is just one example of ways that communities can come together to be self sustainable and care for each other as well as the environment.

Another example of a way that community engagement can work comes from a research paper titled, "Flexibility & Structure: Community Engagement on Climate Action & Large Infrastructure Delivery" by Evan Boyle, Martin Galvin, Alexandra Reves, Aoife Deane, and Brian Ó Gallachóir. This paper looks into the role of citizen and community engagement in Ireland and discusses policies and frameworks that can be used to develop climate related developments. When talking about Ireland's climate action plan and how to implement it they state, "Recent policy developments, both nationally and at a European level, support the need for a re-evaluation of community engagement within public bodies in order to transform their relationships with citizens/communities in order to achieve the scale of change needed to respond to the challenges posed by climate change." (Boyle, et al. 4). This means that countries need to be thinking about community engagement more than they have in the past. This can be done in a multitude of ways, but the example that was given was that Irish public bodies are working alongside local authorities in order to implement these climate action initiatives. This can be a way that people are getting involved and showing that they care about this issue. From the research that they did, there is a clear structure to gaining community engagement and how policy makers and leaders of different climate organizations can include the public in their work to better the climate.

One aspect of community engagement in climate action is community gardens. Community gardening is a way of using sustainable practices while also being a way for members of the community to connect with one another. In the article, "Scaling Up Community Action for Tackling Climate Change" by Deirdre Shaw, Andrew Cumbers, Robert McMaster, and John Crossan, they talk about the importance of community gardens and how it can be a way of connecting community members to broader sustainability agendas. They state, "For some community gardeners, they just want 'to dig holes in the ground' (Digley Diggers, community garden volunteer), while others wish to re-imagine the city through the use of existing derelict sites and peri-urban spaces to promote sustainable food production and consumption, with the attendant environmental benefits" (Shaw, et al. 270). This shows that the reasoning for why people participate in gardening can differ, but the point is that people are willing to do it and it can be one small step towards employing sustainable practices that benefit the environment. Community gardens are just one of the ways that community members can participate in climate action.

In the same article, they talk about the importance of community gardens in urban areas and places where air pollution often affects the health of those living in larger cities. They state that sustainable practices like community gardens can improve the geographical environment. They go on to state, "Community gardening and similar activities reduce 'heat island' effects by, for example, increasing vegetation density in urban areas, improving waste management through composting, simplifying supply chains by reducing 'food miles' and transport flows through local food production and, hence, contributing to sustainable climate change initiatives" (Shaw, et al. 267). This shows that community gardens are helpful for many reasons, both for the people that will benefit from the food and improved air quality, and for the communities that will thrive from participating in community initiatives. They go on to state that, "The role of community gardens in promoting social inclusion and community building has been well documented (e.g. Crossan, et al., 2016). Prugh, Costanza and Daly (2000) argue that community is central to sustainability, providing opportunities for social empowerment at the local scale of everyday life." (Shaw, et al. 271). This goes to show that community is an essential part of climate action and that when talking about climate change, people should be focusing on how people can come together to participate and to make communities operate sustainably.

Another example of climate action that can be implemented is ecovillages. This term is explained in the article, "The Paradox of Collective Climate Action in Rural U.S. Ecovillages: Ethnographic Reflections and Perspectives" by Chelsea Schelly, Zach Rubin, and Joshua Lockyer. They state that, "Ecovillages involve forms of collective dwelling where members adopt alternative ways of living in active contrast to mainstream social systems, by building alternative dwellings9, approaching technology through an alternative definition of modernism11, or adopting an ethos of degrowth rather than uncontrolled capitalist growth" (Schelly, et al. 2024). This form of climate action is more involved than others but is something that shows the willingness to participate when people are in a community with similar values. They go on to talk about the lives of those in ecovillages and state that, "Often, ecovillage residents promote their way of living as more ecologically responsible than mainstream ideas, mindsets, ways of being, and consumption habits when it comes to taking climate action" (Schelly, et al. 2024). These ecovillages can be a way of getting people who have similar lifestyles and who care for the environment together to use sustainable practices and reduce their overall consumption. This can not only be good for the environment, but can be a way that people find community in doing climate action.

There are just some examples of why it is important to change the way that climate change is talked about in mass communication. Aspects of joy and community can be included to get everyone inspired and involved in climate action initiatives. Community gardens are one simple way to get people involved, but there are a multitude of ways to get communities to come together to make a difference. It is important that in doing work to reverse the effects of climate change there is care and intention put into every aspect, including how people care for the individuals in the community. If there is no intention put into climate action initiatives then there is a higher chance that individuals will burn out. To ensure that those participating in climate action stay resilient in their efforts, then there should be more of a focus on joy and community. Community engagement as a climate action tactic because people are more willing to work in groups than to work alone. That is why there is a need to adjust how environmental communication is handled when it comes to climate change.

## References

Boyle, E., Galvin, M., Revez, A., Deane, A., Ó Gallachóir, B., & Mullally, G. (2022). Flexibility & structure: Community Engagement on Climate Action & Large Infrastructure Delivery. *Energy Policy*, *167*, 113050. https://doi.org/10.1016/j.enpol.2022.113050

Craps, S., & Olsen, I.M. (2020). Grief as a Doorway to Love: An Interview with Chris Jordan. American Imago 77(1), 109-135. https://doi.org/10.1353/aim.2020.0006.

Johnson, A. E. (2022). How to find joy in climate action [Video]. In *TED Talks*. https://www.ted.com/talks/ayana_elizabeth_johnson_how_to_find_joy_in_climate_action/transcript.

Parshley, L. (2022, July 6). Happy Climate is on a mission to show what we gain from climate action. Fix. https://grist.org/fix/advocacy/happy-climate-project-reduce-carbon-footprint-find-joy/.

Salvador Costa, M., Leitão, A., Silva, R., Monteiro, V., & Melo, P. (2022). Climate

Change Prevention Through Community Actions and Empowerment: A Scoping Review. *International Journal of Environmental Research and Public Health*, *19*(22), 14645. https://doi.org/10.3390/ijerph192214645

Schelly, C., Rubin, Z., & Lockyer, J. (2024). The paradox of collective climate action in rural U.S. ecovillages: Ethnographic reflections and perspectives. *Npj Climate Action*, *3*(1), 1–9. https://doi.org/10.1038/s44168-024-00100-w

Shaw, D., Cumbers, A., McMaster, R., & Crossan, J. (2018). Scaling Up Community Action for Tackling Climate Change. *British Journal of Management*, *29*(2), 266–278. https://doi.org/10.1111/1467-8551.12274

## Learning Outcomes in Communication

LOC #3: Engage in Communication Inquiry
LOC #4: Create Messages Appropriate to the Audience, Purpose, and Context
LOC #9: Influence Public Discourse

National Communication Association. (n.d.). https://www.natcom.org/sites/default/files/publications/LOC_1_What_Should_a_Graduate_with_a_Communication_Degree

IdeaFest
Journal

# Determining the Quality and Quantity of California Sitka Spruce (*Picea Sitchensis*) on Private Commercial Timberland for use in Domestic Acoustic Steel-String Guitars

**Ryan Cooper**


## Abstract

Development of additional revenue streams for timber products with relatively low profitability is an important step in securing future economic success for the forestry sector in Humboldt county. Utilization of California Sitka spruce (*Picea sitchensis*) for use in domestic acoustic steel-string guitar manufacturing may provide exactly that. Determining the quality and quantity of California Sitka spruce is the logical first step in achieving this.

Quality was determined from tree core samples gathered at sites on Green Diamond Resource Company's (GDRco) California timberlands and the Arcata Community Forest (ACF) with high basal area of Sitka spruce. Rings Per Inch (RPI) is the most significant indicator in quality for guitar manufacturing (Pedgley et al., 2009) and was used as the anchor for tree core analysis. Cores were analyzed for the mean annual RPI of the stem and other physiological measurements. A minimum mean RPI of 10 was used to determine acceptability for manufacturing use (Stephens, 2015). Statistical tests determined significant differences in mean RPIs between GDRco and the ACF as well as between individual GDRco sample sites. Age was the most significant quantitative predictor of RPIs. The analyses indicated that Sitka spruce in this region is capable of producing the desired RPIs at an optimal age. Quantity requirements were calculated from several sources and results indicated that the region's inventory is sufficient for domestic manufacturing needs.

Management objectives and site-specific silvicultural prescriptions may be key in achieving the desired results in accelerated time frames to meet GDRco management goals (GDRco FMP, 2023).

## Background

The extraction of timber for use in forest products has a long and storied history. Revenue streams that have developed from these markets have solidified around products that have matured through centuries of refinement; and the supply of timber for these products being a viable option for small and large landowners alike (Youngs, 2003). Further refinement of these categories, including but not limited to pulp, dimensional lumber, and fine hardwood and softwood production suitable for instrument manufacturing and furniture manufacturing often means producers having to source raw materials that are either cheaper and or easier to acquire (Gibson and Warren, 2016). The steel-string (and to a lesser extent but still notable) nylon-string acoustic guitar manufacturing industry has long sought to find continually cheaper and/or easier to source materials to support or extend their existing sources. However, as the market has continued to grow over the last century, that task has become increasingly difficult (Taylor, 2018; Gibson and Warren, 2016; Music Trades, 2024).

A significant driver of continued natural resource procurement in the steel-string acoustic guitar manufacturing industry stems from the desire by many manufacturers to continually produce the same models with the same species of wood, often over decades, and sometimes centuries (Bennett, 2016; Gibson and Warren, 2016). This often comes at the expense of sustainability, and is a growing concern industry wide (Gibson and Warren, 2016; Fretboard, 2018). With global concern about a changing climate and a growing call for more sustainable materials and practices (French et al., 2009), altering or changing existing supply sources of specific timber species required for acoustic guitar manufacturing may become a driving force for many guitar manufacturers. There have been many attempts to redefine what the standard acoustic guitar is made of over the last 50 years, some stemming from the desire to create something new, some from the desire to reduce the ecological impact associated with the harvest of timber, and some from the need to provide their customers with a dependable and recognizable product (Calvano et al., 2023). While many of them have proven effective to a certain degree in terms of market share, but they lack the robust and highly desirable sonic properties provided by high quality tonewoood (Pedgley and Norman, 2012) which may prevent them from ever being the first choice for consumers.

In addition to the significant audible differences, global manufacturers and home luthiers alike have a robust commitment to building acoustic guitars in their traditional way and with traditional materials (Bennett, 2016). An important companion to the adherence and use of traditional materials to maintain the continued growth of acoustic guitar sales, are the stringent quality specifications for wood upheld by most manufacturers, especially in higher price point guitars past the $500 retail price range.

A steel-string acoustic guitar is made of many different parts that come from a wide variety of tree species (Figure 1).

With respect to the fact that there are many ways to build an acoustic guitar, the application of solid wood is by far the most important differentiator in determining the sound quality, longevity, and long-term value for acoustic guitars (French et al., 2009). The development of what we commonly know today as the standard acoustic guitar was developed in the early 20th century by C.F.

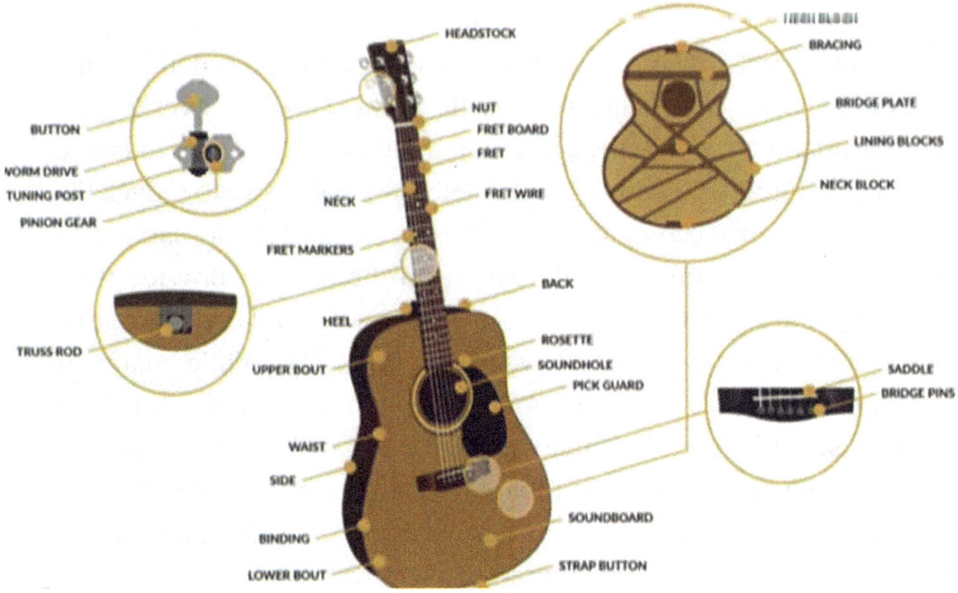


*Figure 1. External and internal views of the steel-string acoustic guitar and the components that comprise its makeup. Source: theacousticguitarist.com*

*Figure 2. C.F. Martin D-18 steel-string acoustic guitar. Source: martinguitar.com*

Martin (Bennett, 2016). It was the final revision of this instrument that has seen many centuries of iterations and revisions previously, finally arriving on what many feel is the perfect combination of sonic materials (Figure 2).

Coupled with a growing economy and the growing appreciation of popular music (Schloss et al., 2012), especially in the United States, its arrival exploded in popularity with steady increase lasting over a century and continuing to this day. Acoustic guitars of this era (early 20th century) were made

exclusively from pieces of solid wood. The top being composed of a quarter-sawn and butterflied solid piece of Sitka spruce (*Picea stichensis*), and to a lesser extent various types of false cedar, typically western red cedar (*Thuja plicata*), while the back and sides of the instrument were typically composed of tropical hardwood. The bracing that lines the underside of the topwood as well as the back piece is also typically made of Sitka spruce. The preferred back and side wood was, and still is, rosewood (*Dalbergia spp.*). The sourcing practices of rosewood coupled with the guitars meteoric rise, lead to exploitation of the materials by manufacturing from the guitar industry and many others. By the 1960s, international bans and restrictions from export on the most desirable type of rosewood Brazilian rosewood (*Dalbergia nigra*) were set in place (Arbelaez, 2022).

The export ban of Brazilian rosewood led to a shift in global sourcing for rosewood with new locations around the globe appearing and disappearing as the product was overharvested. A contemporary example of this may be found in Madagascar, where rosewood has been banned from export. It should be noted that there was a brief one year exemption to the ban in 2009, which manufacturing took advantage of, but only furthered the illegality of the source as exporters (illegal and legal alike) were able to claim for multiple years following the exception that their product was harvested during the exemption year (Innes, 2010). The most common source for rosewood following the Brazilian rosewood ban in today's current markets is India, which supplies the market with east Indian rosewood (*Dalbergia latifolia*). Mahogany (*Swietenia macrophylla, mahagoni, humilis*) is the second most popular tonewood species that is utilized in significant quantities for acoustic guitar manufacturing and is also facing increasing scarcity and exploitation (Martinez-Reyes, 2015). Tonewoods that comprise the back and sides of guitars usually serve three purposes; they play a role in determining how the ultimate sonic profile of the guitar will sound, they are important in structural integrity, and they add signifacnt visual appeal.

Acoustic guitars produce sound by vibrations of plucked or strummed strings under high tension (up to 180 psi) that vibrate through the top of the guitar and resonate against the back and sides of the instrument in the acoustic chamber. The string vibrations are transferred to the top of the guitar via the bridge which agitates the top, and captures the string vibrations and produces its own sonic waves both longitudinally and radially. Much of what humans are able to audibly perceive is derived from these longitudinal and radial wave propagations (Merchel, 2019). The egress of air from inside the body of the guitar is what we ultimately perceive as acoustic sound. In comparison to the back and side woods, the wood comprising the top of the guitar is by far the most significant in producing desirable sonic properties.

While it is important to highlight that all three have a part in producing sound wave propagation, the topwood is the singular focal point for the production of soundwaves. Centuries of manufacturing acoustic instruments such as violins and pianos have all arrived at a similar conclusion, spruce (*Picea spp.*) is the best genus globally to use when creating a soundboard (Bennett, 2016). Of the many varieties of spruce Sitka, is one of the most robust in terms of stiffness and strength and also has one of the highest strengths to weight ratios of any softwood (Gore, 2011) making it ideal for use as a soundboard in acoustic guitar manufacturing. It is also readily available in a multitude of locations globally, making it the preferred species for steel-string acoustic guitar manufacturing since the instrument's inception (Gore, 2011). Although the Sitka is on a volumetric scale, plentiful, the specific requirements for instrument-grade Sitka are not. Manufacturers and luthiers alike seek raw materials that are derived from stems with very minimal annual diameter growth which produces tightly spaced annual growth rings, known colloquially as tight-grain. In addition to these raw material requirements, manufacturers also need stems with enough diameter to make their products. These requirements result in the need for the oldest Sitkas within any stand that they are found (French et al., 2009).

## Introduction

Continued and growing demand over vast temporal scales often induces significant strain on supply for any industry that relies on natural resources that take decades to reach maturation. This has been especially true for acoustic guitar

*Table 1. Ranges and preferences of musical instrument grade Sitka Spruce tonewood top and brace billets. Data sourced from: Gore, 2011, Merhel et al., 2019.*

| Density range ($lb/cf^3$) | Density preferred ($lb/cf^3$) | Young's Elastic Modulus (Gpa) | Elastic Modulus preferred (Gpa) | Mean Janka Hardness. ($lb_f$) |
|---|---|---|---|---|
| 22.8 - 26.5 | 22.8 | 11.1 - 15.7 | 12.2 | 472.5 |

manufacturers (French et al., 2009). Although wood products derived from trees are, on a basic level, sustainable, the cost of continually sourcing almost indistinguishable raw material (raw logs) can be significant (Calvano et al., 2023). Cost increases in extraction, transportation, and processing are well known throughout the forestry sector (Haynes, 2003). This is further compounded by the fact that the materials required and sought for steel-string acoustic guitar manufacturing are traditionally the oldest and largest stems within the forest matrix that they are found in.

The physiology of spruce makes it highly adept not only at performing the function of a soundboard for any acoustic instrument but also in its volumetric abundance both regionally and globally. Spruce's density to weight ratio allows for a product that is light enough to be worked and held while also providing the structural integrity required for guitar manufacturing (Table 1). These physiological properties that allow spruce to be minimal in weight while also providing a significant amount of structural integrity, the key and desired component of steel string acoustic guitar manufacturing due to the tension from steel strings on an acoustic guitar resting at 180psi when tuned to pitch.

supporting the overall structure of the guitar, spruce's ability to propagate soundwaves both longitudinally and radially are what truly allow the species to be regarded as the best sonically (French et al., 2009). The way in which spruce, or any organic/inorganic substance, exhibits dynamic mechanical properties inherent to the creation of sound waves is referred to as Acoustic Theory. In short, acoustical theory examines the speed (*C*) at which sound waves are produced. It is dependent on the density of a particular substance (*p*) and its modulus of elasticity ($E_L$). The equation can be simplified to:

$$\mathbb{C} = \sqrt{\frac{E_L}{\rho}}$$

Density is derived from the dried mass of a particular substance, and in our case we will be examining Sitka spruce. Youngs modulus of elasticity is derived from a mechanical test of pressure over strain. It examines how much pressure can be exerted both longitudinally and radially over a given sample length before the sample exhibits plastic qualities and begins to deform. This measurement is referred to as the elastic modulus of a particular substance. When looking back at our equation (equation figure), *C* is the resulting speed at which waves are propagated. Derived from the modulus of elasticity and the density.

There is significant published literature to rely on in determining the average density for instrument grade spruce topwood, and indeed there is also recent literature available for elastic modulus testing regarding the same subject (Merchel et al., 2019; Badiane et al., 2023). The unique way humans interpret sound, will no doubt alter one individual's choice of the most pleasing sound as compared to another, yet the significant amount of available literature allows us to arrive at a very defined range average that allows for precise measurements when selecting raw materials (Table 1).

While spruce as a general species provides the basic needs for acoustic tone production, there are significant growth trait requirements that are key components in determining the viability of an individual stem for acoustic guitar production. Annual growth ring production, including both earlywood and latewood production and their overall radial width in comparison to the overall diameter of the individual stem are all significant. Manufacturers refer to this calculation as Rings Per Inch (RPI). A low RPI measurement significantly affects the elastic modulus of spruce, which not only reduces its structural integrity but also, too few RPIs often produces an unrefined sound (French et al., 2009). While there is no defined maximum, having a mean of 20+ RPI across the width of the topwood often causes the top of the guitar to look plastic at a distance and is undesirable, especially in upper price point guitars (Figure 3). Generally speaking, a mean of 10 RPI or above produces a highly desirable tone and from a visual standpoint, 10-20 RPI is preferred, with 15-25 being the most desirable (Stephens, 2015; Ryan, Ryan Guitars; Haines, Guild Guitars; Figure 3).

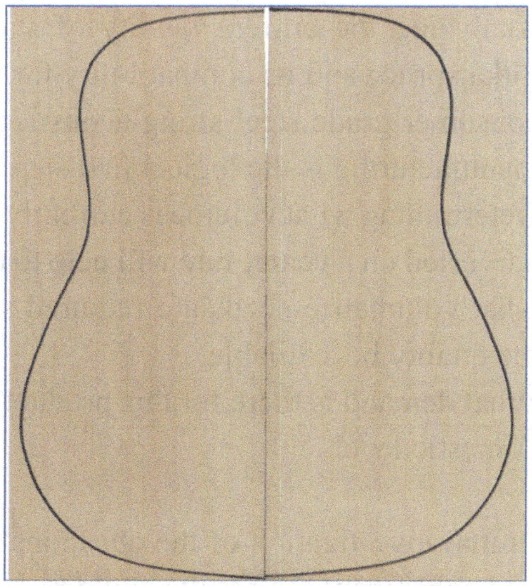

*Figure 3. Sitka spruce steel-string acoustic guitar top blank of desirable quality with 16 mean Rings Per Inch (RPI). Source: Maderas Barber Tonewood*

Considering the physiological make up first, the visual component of an acoustic guitar is a very important and close second in determining its selection for use in acoustic guitar manufacturing. These qualifications have almost exclusively been derived from specimens that tend to be slow growing and exceed 200 years in age (French et al., 2009). Raw materials of this quality were easily acquired through the early to late 20th century. Sitka spruce was utilized for a number of various purposes other than instrument manufacturing in the early to mid-20th century, all of which contributed to the decline of fine timbers. Outlets for the material ranged from airplane manufacturing, requiring intense and specific standards not unlike those for guitars, down to pulp production, requiring very little in regards to fineness and quality. This subsequently led to the over exploitation of mature spruce found within the United States, and by the post-war mid-20th century, yearly MMBF (million board feet) production was in decline from the largest producing ports. There was a tremendous boost to yearly production through the late 80 and 90s by a magnitude of five, but that has since dwindled to a level matching those of the mid post-war 20th century (Brackley, Haynes, 2008).

It should also be noted that part of this decline was also in response to increased environmental regulation, and the opening of new manufacturing sources from Asia (Gibson and Warren, 2016). Whether the focus is on the source for top or back and side tonewoods, the stringent requirements will typically arrive at the same thing, materials derived from mature or decadent trees (French et al., 2006). Many of the locations where these species are obtained have a history of extraction covering more than a century, and trees with the desired physiological traits are becoming increasingly more difficult to secure.

Despite several non-contiguous years of negative growth and the economic downturn associated with the Covid-19 pandemic, the steel string acoustic guitar has seen its popularity rise continually from 2013 - 2022 (the most recent year sales data was available). In that time frame, the acoustic guitar market has seen its domestic retail value rise from $603,195,000 to $1,041,900,000, a 73% increase over the 10-year period. Wholesale value has risen from $404,141,000 to $666,816,000, a 65% increase (Music Trades, 2023). This supports the assumption that while global music tastes will change including that of the general consumer for these products, the continued growth and popularity of the acoustic guitar will change very little and as such neither will the demand for its raw materials.

## Objectives

With such abundance of this highly prized species (spruce) growing in easily accessible areas within northern coastal California, there is a notable lack of supply coming from the region for use in guitar manufacturing. Manufactures are looking continually for new sourcing locations that meet their quality demands, are sustainable, provide a lighter carbon footprint and offer economic incentive to local economies. Not only to strengthen their supply chains but also to provide the retail end-user with the highest integrity product available.

Northern coastal California has the potential to supply the needs of the domestic acoustic guitar manufacturing industry with Picea sitchensis (Sitka spruce) in perpetuity. Although significant and long running research, as well as management plans that meet specific needs of both landowners and manufactures is lacking, there is abundant data, tree volume, and public appetite to warrant further research. These assumptions form the basis for the objectives of this thesis and are as follows:

1. Analyzing the current quality of standing Sitka spruce and its compatibility for use in consumer grade steel-string acoustic guitar manufacturing is the logical first step.
2. Determining what volume is currently being harvested on a yearly rate will help to define what volumetric needs are required should the quality be available.
3. What demand is there for this product at the domestic level.

The initial investigation of the questions posed above and accompanying results could be used as a framework for developing a new revenue stream in northern coastal California for commercial and small timberland owners alike.

## Methodology

### Site Description

All sampled units, with the exception of those originating from the Olympic Peninsula, were collected in Humboldt County, and approximately 2.25 miles or less to the coast line. The area is classified as a warm temperate rainforest (Waring and Major, 1964) and receives a majority of its rain fall from October to April with a mean from all sample sites of 48.25". The sample locations of Big Lagoon, Murray Road, and Eureka all occurred on Green Diamond Resource Company (GDRco) property. The Murray Road and Eureka locations

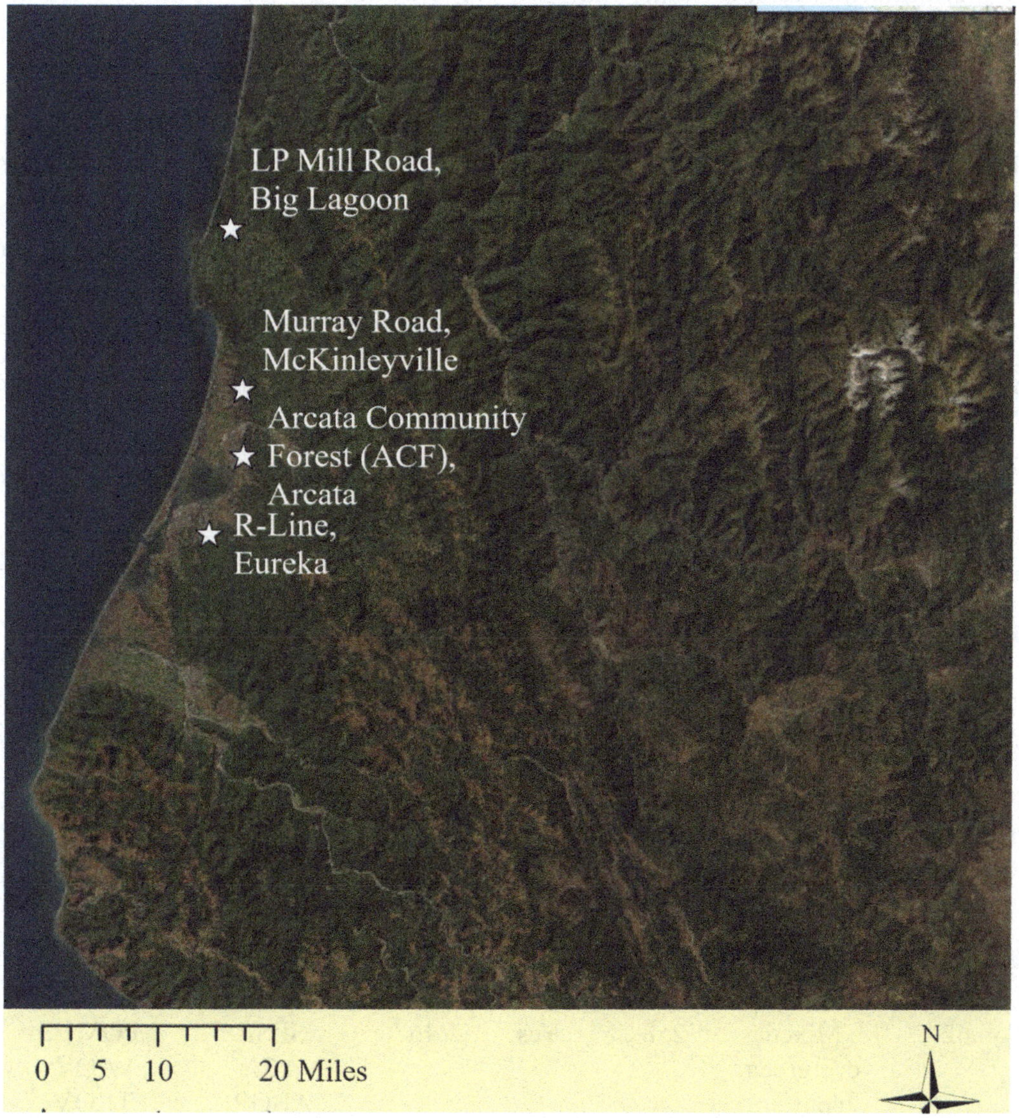


*Figure 4. Map of coastal Humboldt County, located in northern California, showing locations of all sample areas. Samples originating from OP are omitted from this map.*

share boarders with the McKinleyville community forest and the McKay community forest, respectively. The sample locations at the Arcata Community Forest (ACF) occur in a community forest that is managed by the city of Arcata for multiple recreational purposes, including management to improve forest health and increase standing volume (ACF FMP, 2020). GDRco utilizes an evenaged regeneration method while the ACF operates under an uneven aged silvicultural system. Additional information and details about each sample location may be viewed in Table 2, and a location map is provided below (Figure 4).

Five of the samples originating from the ACF were not collected in the field as others were. The samples were accessed at the Kenneth L. Fisher Laboratory of Redwood Forest Ecology at Cal Poly Humboldt (Dendrochronology lab). Those samples were collected as part of a separate study not affiliated with this project. Samples were collected by Stephen C. Sillett, and additional information about them may be found within the published paper "Comparative development of the four tallest conifer species" by Sillett and others (2021). Olympic Peninsula (OP) data originated from samples also stored at the Dendrochronology lab. Samples were collected by Russell D. Kramer and others, and were used as part of a separate study. Further information regarding those samples may be found by viewing "Quantifying aboveground components of *Picea stichensis* for allometic comparisons among tall conifers in North American rainforests" by Kramer and others (2018).

*Table 2. General site description details for all sites sampled for tonewood. Samples were taken at four distinct locations along the coast of Humboldt county*

| Site Description | Mean Elevation | Forest Type | Soil Type | Public Access | Yearly Rainfall | Overstory | Understory | Site Class | Recent Entry |
|---|---|---|---|---|---|---|---|---|---|
| Big Lagoon | 287 | Mixed evergreen conifer 232 | 257,258, 590 | No | 53" | SESE, PSI, ABGR, PSME | POMU, VAOV, RUUR RUAM | I & II | Yes |
| Murray Road | 202 | Mixed evergreen conifer 232 | 258 | Yes | 50" | SESE, PSI, ABGR, PSME | POMU, VAOV, COSE | I & II | No |
| Eureka | 194 | Mixed evergreen conifer 232 | 397,398 | No | 44" | SESE, PSI, ABGR | POMU, VAOV, RUUR RUAM | I & II | No |
| Arcata Community Forest | 426 | Mixed evergreen conifer 232 | 258 | Yes | 46" | SESE, PSI, ABGR, PSME | POMU, VAOV, TROV | II | No |

## Stand Description

Data provided by internal inventories conducted by Green Diamon Resource Company (GDRco) determined that throughout their current ownership, 122 stands exist with >50% basal area per acre (BA) of spruce, with a total area of 1,565 acres. The delineation of >50%BA Spruce was chosen as those stands present the highest likelihood of locating individual stems with greater than 20" DBH, which is the minimum requirement for production of tonewoods (French et al., 2009; Gibson and Warren, 2016) While throughout the entirety of GDRco ownership spruce is found in areas totaling to 29,314 acres, time constraints for collecting field data required a focus on the stands with the greatest BA of Sitka spruce.

## Sample Design

Of the 122 stands, 12 were selected for field survey. Final selection was determined by analysis of the supplied inventory data for the project and GDRco internal GIS layers. The inventory was analyzed and manipulated to show only the species of interest (Sitka spruce) and was further refined by DBH class. Stands that contained significant portions of small diameter class spruce (<18" DBH) were rejected from additional analysis. Further consideration was given for time constraints for collection of field data (e.g. proximity to the college and vehicle access to the individual stands), in making the final determination for which stands would be sampled.

Lastly, consideration for stands to be examined was provided by professional opinion from GDRco employees that were familiar with local conditions and historical management of the stands in question. The final stands that were determined to have advantageous locations for sampling the individual physiology of spruce trees were termed *candidate stands*. Candidate stands were subdivided into unique strata using logical geological features such as ridge tops, riparian areas, roads, and again for available time in the field. The target sample set of candidate stands totaled to approximately 4% off the 122 stands with >50% BA spruce. Confidential inventory data provided by GDRco contained stand locations as well as individual tree measurements including DBH and height. Due to robust inventory data, a simple transect survey method of the candidate stands was performed to locate the cruised trees as well as any others that fell outside of the original cruise. Transect surveys were conducted within the pre-selected candidate stands using the candidate stand cruise boundary as starting locations.

## Sample Method

Straight line transects were established in the field and were determined based on visual analysis of each unit that was visited. Considerations for geologically unstable areas, lack of mature overstory spruce stems and recent completion of final harvest. An azimuth reading was taken to determine the direction of travel and to ensure that more than one sample stem would be hit. While it was important to consider the scientific standard of sampling protocol, the time constraints of the project also justified additional considerations as to where the transect would be conducted. In addition to a handheld GPS unit, Avenza maps was used to mark locations. Avenza was also used to provide transects boundaries. Preliminary transects boundaries were added to the map prior to field research and were adjusted accordingly while gathering field data. Visual determination of >20" DBH stems was easily achievable and further instrumentation to confirm this was unnecessary. Due to the short window available for data acquisition, stems at the onset, middle, and termination of the transects were chosen for sampling.

If DBH met or exceeded 20", additional measurements were taken and included the height of the individual tree. The live crown ratio as well as the length of clearwood present on the bole. Any unique features of trees were also noted including but not limited to: crook, sweep, broken top, signs of stress or damage from pests or disease. Canopy class within stands was defined as dominant (D), co-dominate (CD), intermediate (I), and suppressed (S). The tree was also cored using a Haglof increment bore of appropriate size to ensure the pith was hit. The core was measured using a ruler with 1/10th inch increments.

Regardless of initial reporting of rings per inch, all sampled stems were mark with three dots around candidate tree at breast height, as well as a butt mark of SR (spruce research) painted on the downhill side at the basal area. Yellow marking paint was used for all marks.

Tree location was marked with the GPS unit and/or Avenza. The tree core was stored in a plastic straw and returned to the laboratory for further analysis. All collected measurements and notes were recorded in field notes to then be compiled and refined during data analysis. Due to time constraints, additional measurements and analysis of tonewood quality, such as dry density and modulus of elasticity were beyond the scope of this project.

Additionally, core samples were also collected from the Arcata Community Forest (AFC), to provide additional context into differing growth patterns across several different management objectives and silvicultural prescriptions. Samples collected from the ACF were collected following the same protocols listed in section 3.4.

## Provided Samples

Prepared samples were also examined at the Kenneth L. Fisher Laboratory of Redwood Forest Ecology at Cal Poly Humboldt (Dendrochronology lab). Prepared samples provided by the lab were selected for their collection location, which included both the ACF as well as locations from Mt. Rainier National Park located in the Olympic Peninsula (OP). All selected samples were chosen to highlight the difference in growth patterns exhibited by various management schemes. Measurements taken from both the fresh samples as well as the historical samples at the Dendrochronology lab included RPI and total age of stem. Measurements were used to compare against samples taken from GDRco. ownership sample units.

## Tools

The following tools were used to gather the required data.

- Flagging (orange and blue?) - 3 points around candidate trees >20" DBH
- Nikon Forestry Pro II and/or TruPluse Laser hypsometer
- 32" & 18" Haglof Increment borer
- Biltmore stick
- Suunto Mc2 Compass

- Spencers Logger Diameter Tape - 75'
- Suunto PM-5 Clinometer
- Survey tape: Cloth 100' open-reel tape
- Field notebook & pencil or pen
- Talley sheets
- 1/10th inch increment ruler
- Plastic straw to store tree cores
- Garmin GPS 64st
- Yellow Marking Paint

## Domestic Steel-string Acoustic Guitar Sales Data

Data to determine the quantity of acoustic guitar instruments sold on a global scale annually was acquired from Music Trades. Retail and manufacturing data was analyzed from the most current year available (2022) as well as the previous five years. The data package was refined by removing all other guitar and instrument sales (electric guitars, other acoustic instruments such as ukulele and mandolins etc.).

We also chose to use volume measurements of brace wood and soundboard billets that a large volume material supplier to global acoustic guitar builders used for additional depth to annual consumption of raw material (Merchel et al., 2019). Elastic modulus testing and dry density testing required significant additional time investments and conducting our own testing for these was outside the scope of this project. Reference data for this portion of wood dynamics evaluation was obtained from several studies that have examined dried, cured, and manufacturer ready samples of Sitka spruce (Merchel et al., 2019). Janka hardness ratings were gathered from multiple literary sources (Gore, 2011, Merchel et al., 2019). Basic conversions were used to calculate the required cubic meters, cubic feet, and board feet required for brace wood and topwood billet production. Specific volume measurements were gathered from an "all-in-one" acoustic guitar builders kit available at www.stumac.com and was then extrapolated to infer the annual requirement to supply the global acoustic guitar market.

## Quantity of Spruce

Determination for the quantity of available spruce that could potentially be delivered on a yearly basis and in what classification was accomplished through several processes. Confidential delivered raw log data was acquired from GDRco from the years 2016-2020. Data contained all sizes of harvested and delivered logs ranging from 6" small-end diameter (SED) up to 54" SED as well as total log lengths. Data was manipulated to omit undesirable sizes classes and present only sizes classes of interest to the current research project, leaving SED sizes classes ranging from 20"-54". Further refinement was accomplished by refining the data down into yearly outputs and converting board feet production into cubic feet.

Data gathered from the field was compiled and numerous statistical tests were performed using excel and MiniTab. Tests were performed to determine if there was any significant relationship between the individual physiology data sample points their ability to produce material that is required by global acoustic guitar manufacturers. Information on what order of importance determines the utilization of logs from Sitka spruce trees was determined by analysis of current published literature as well as information provided by individual luthiers.

*Table 3*. Sample summary data for all continuous and categorical variables.

| | Big Lagoon | Murray | Eureka | ACF | OP |
|---|---|---|---|---|---|
| Mean RPI | 3.4 | 3.6 | 5.6 | 10.5 | 12.9 |
| RPI Range | 2 - 5 | 2 - 5.5 | 4 - 8 | 7.5 - 14.3 | 8.7 - 15.5 |
| Mean DBH (in) | 30.9 | 25.4 | 22.4 | 62.1 | 98.6 |
| DBH Range (in) | 21.4 - 39.3 | 18.1 - 35.3 | 18.9 - 26.6 | 22 - 130.7 | 75.3 - 125.4 |
| Mean Total Height (ft) | 144.7 | 146.2 | 106.8 | 198.7 | 274.4 |
| Total Height Range (ft) | 106.8 - 171.1 | 99.1 - 167.2 | 75.1 - 118.3 | 128.8 - 284.1 | 257.9 - 295.7 |
| Mean Clearwood (ft) | 23.6 | 22.7 | 42.8 | 73.3 | - |
| Clearwood Range (ft) | 0 - 63 | 8 - 53 | 32 - 60 | 74 - 89 | - |
| Mean Crown Ratio (%) | 69 | 66 | 38 | 40 | - |
| Crown Ratio Range (%) | 40 - 90 | 30 - 90 | 20 - 40 | 30 - 50 | - |
| Avg Crown Class | CD | CD | CD | CD | - |
| Crown Class Range | D - CD | D - I | CD - I | CD - I | - |
| Mean Age | 40 | 38 | 38 | 165 | 307 |
| Age Range | 38 - 49 | 32 - 45 | 27 - 44 | 89 - 318 | 256 - 353 |

## Results and Discussion

### Full Data Set Quality

A regression model was constructed and analyzed utilizing all logical variables (Table 3). The analysis that was preformed resulted in significant outcomes for the continuous predictor variable of Age and categorical variable Region Number. An interaction term was built to analyze the relationship between DBH and Age within the model as earlier versions showed significant results between these two predictors. The final model increased the *P-Value* for DBH slightly above significant levels but the Age remained unchanged. The results also indicated that the collected data fit well with the model with a final $R^2$ adj value of 88.85% (Table 4). Further Analysis was applied to the mean RPIs across all collected samples and Region Number. The results indicated there was not a significant difference in RPI between sample locations originating from GDRco ownership, but did report a significant difference between GDRco and ACF (Table 6). Significance was even strong when analyzing GDRco mean RPIs and those originating from the OP but those results were omitted due to the OP location falling outside of the study area.

A second regression model was created to analysis the predictability of RPI as the response to age for all samples. The model reported strong fit as well as a significant relationship between the predictor, age, and the response, RPI (Figure 5). RPI means for all samples were then analyzed to discern any significant differences between them. The mean RPI was lowest in Big Lagoon and largest in the Olympic Peninsula (Figure 6). Following this analysis, a two-sample T-Test was performed to discern any significance between the results of the mean RPIs from all GDRco samples

*Table 4. Regression model variance and model summary results for all samples across all Regions.*

| Source | *DF* | *Adj SS* | *Adj MS* | *F-Value* | *P-Value* |
|---|---|---|---|---|---|
| Regression | 8 | 619.26 | 77.4075 | 41.84 | 0 |
| DBH | 1 | 4.848 | 4.8484 | 2.62 | 0.115 |
| Total Height | 1 | 0.904 | 0.904 | 0.49 | 0.489 |
| Age | 1 | 25.621 | 25.621 | 13.85 | 0.001 |
| DBH x AGE | 1 | 1.12 | 1.1195 | 0.61 | 0.442 |
| Region Number | 4 | 31.294 | 7.8234 | 4.23 | 0.007 |
| Error | 33 | 61.058 | 1.8503 | | |
| Total | 41 | 680.318 | | | |

| ***SD*** | ***R**-sq* | **R**-sq(adj) |
|---|---|---|
| 1.36024 | 91.03% | 88.85% |

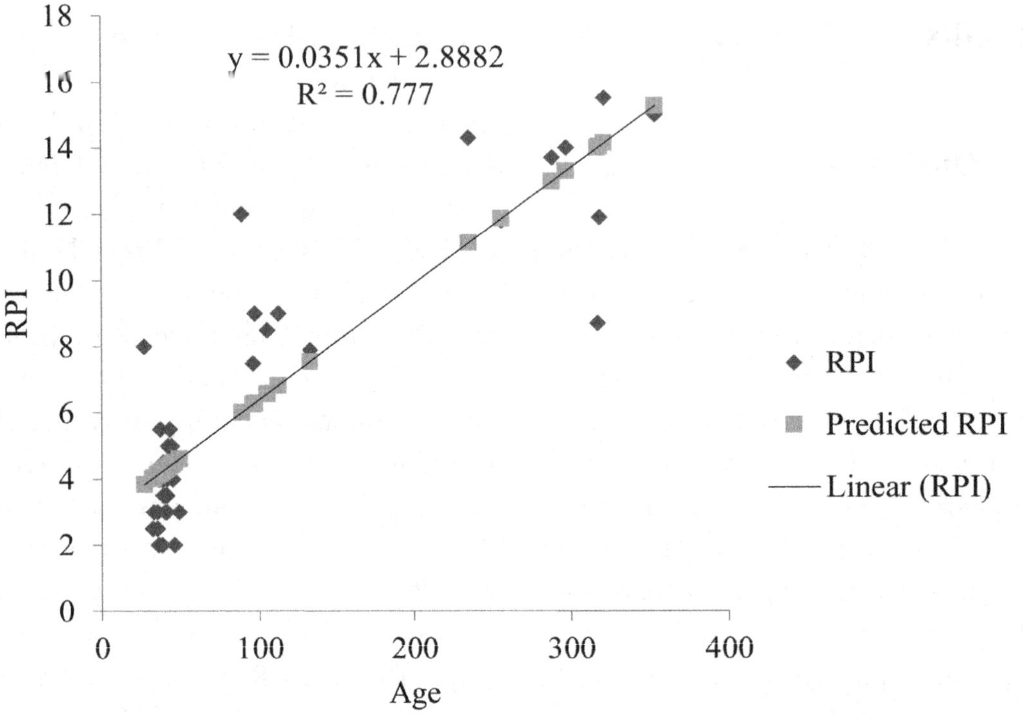


*Figure 5. Mean RPI for each sample from each sample location and total age.*

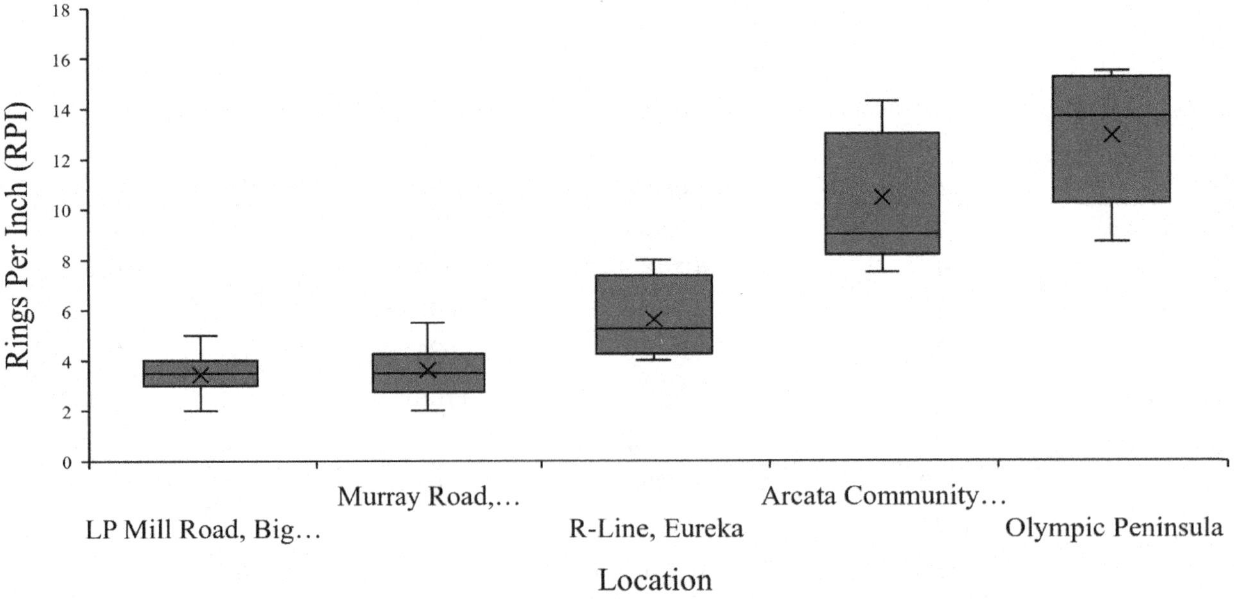


*Figure 6. Mean RPI of each sample location. GDRco samples reported a significant difference in mean RPI than those originating from the ACF and the OP*

*Table 5. Two-sample T-Test test results of GDRco RPI and ACF RPI. The sample size difference was notable between the two but still resulted in significant differentiation.*

| T-Test | Mean RPI | n | SD | SE Mean | T-value | P-value |
|---|---|---|---|---|---|---|
| GDRco | 3.80 | 28 | 1.280 | 0.240 | -7.33 | >0.000 |
| ACF | 10.46 | 9 | 2.630 | 0.880 | | |

and those of originating from the ACF. The results indicated a significant difference in the variability of RPI between the two ownerships (Table 5).

The analysis of all samples from the regression model highlighted the correlation between all individual physiological growth traits measured, except for height. This may only be a byproduct of a limited sample size and bolstering the analysis with additional samples may yield significant results. Further, the significant difference in mean RPIs between the GDRco samples and those from the ACF add strength to the assumption that high quality spruce is located within the north coast region (Figure 2, Table 3). It does however signal that while it is indeed possible to produce high quality spruce locally, aligning management objectives and silvicultural prescriptions are foundational in achieving the desired results.

Taken in totality the analysis indicates that it is important to control growth both on the individual stem level as well as the stand level. There may be no short-term solution for GDRco to reach the required RPI and DBH minimums within their California ownership with current management practices. Examining the mean age of GDRco samples to those from the ACF, 39 and 165 (Table 3) respectively, the ability to produce the required RPIs, while also providing minimum merchantable diameter requirements may only be possible with a much longer rotation age for the selected stems or stands. This would allow the stems to increase in height while restricting diameter growth. This strategy can be further accelerated by reducing crown ratio and keeping the stands in high density for as long as possible and there by producing much smaller annual ring widths (Ashton, Kelty, 2018). Given the short rotation age GDRco employs of approximately 50 years (GDRco FMP, 2023), changes would need to be made in their approach to regenerative evenage methods involving Sitka spruce should they chose to pursue instrument grade spruce tonewood production at the stem level as an additional revenue stream.

It is worth noting that RPIs from additional core samples not included in this analysis examined at the Dendrochronology lab tended to increase as the sample location on the individual stem also increased in height as did the annual ring uniformity. Meaning as you move further up an individual tree the RPI increase and become more uniform. All samples included in this analysis that contained sets of samples at various vertical height intervals all displayed this trait. Sitka spruce originating from GDRco sample locations produced large annual ring widths (sampled at DBH) due to its regional location, site index and high water availability both through traditional soil root interaction but also foliar uptake (Ashton, Kelty, 2018, Kerhoulas, 2020). Without additional samples originating from GDRco locations taken from higher vertical positions along the stem to examine, a comparison is unavailable at this time.

Several possible solutions exist for GDRco to understand RPIs at different positions of the individual stem. First, gathering additional cores from the sample trees at higher positions along the bole would offer the most direct approach to producing additional robust data sets. Second, the proposed analysis may be accomplished by observing harvested stems at various GDRco's log sort locations. Unfortunately, that analysis was beyond the scope of this thesis due to time constraints. The different ownership approaches to silvicultural management and how that affected RPIs aids in understanding what may be possible with alteration to GDRco's current forest management objectives (Figure 6).

## GDRco Quality

A General Linear Model was created to analyze all GDRco samples. The model produced no significant relationships between any of the continuous or the categorical predictors (Table 5). The model explained the data set fairly well but was not as strong as assumed. Further analysis was

*Table 6. Results of General Linear Model for all GDRco sample locations for all independent and categorical variables.*

| Source | *DF* | *Adj SS* | *Adj MS* | *F-Value* | *P-Value* | Model $R^2$ adj (%) |
|---|---|---|---|---|---|---|
| GDRco Unit Category | 1 | 1.5012 | 1.5012 | 2.51 | 0.148 | |
| GDRco DBH | 1 | 1.0287 | 1.0287 | 1.72 | 0.222 | |
| GDRco Height | 1 | 0.5208 | 0.5208 | 0.87 | 0.375 | 63.43 |
| GDRco Age | 15 | 23.3294 | 1.5553 | 2.6 | 0.075 | |
| Error | 9 | 5.3847 | 0.5983 | - | - | |
| Total | 27 | 44.1696 | - | - | - | |

*Table 7. Results of ANOVA analyzing and significant difference between GDRco sample locations of Big Lagoon, Murray Road, and Eureka.*

| Source | *Mean RPI* | *n* | *SS* | *F* | *P-value* |
|---|---|---|---|---|---|
| Big Lagoon | 3.4 | 15 | | | |
| Murray Road | 3.6 | 9 | 15.659 | 6.866 | 0.004 |
| Eureka | 5.6 | 4 | | | |

performed to discern any significance between the Big Lagoon, Murray Road and Eureka RPIs and results yielded were significant (Table 6). Further analysis to gain understanding in RPIs and their relationship to their locations was carried out via an ANOVA Tukey test. The Tukey test resulted in no significant differences between Big Lagoon and Murray, but there was significant difference between Eureka and RPIs from Big Lagoon and Murray road.

The results of the analysis of the samples originating from GDRco ownership did not produce encouraging results for expanding short-term efforts to establish an additional revenue stream for tonewood production. However, the results do seem to support the conclusion that age alone may be the strongest predictor as to the number of RPIs the stem is capable of producing, followed by height and lastly by DBH. While this indicates that achieving the minimum RPI would require changes in silvicultural management (GDRco, FMP, 2020), reaching the minimum is not unattainable. Additional research and study is needed to gain further understanding into what temporal requirement may be required to meet RPI minimums for tonewood production as well as required volume. Altering large scale management plans may be outside of the realm of possibility for GDRco but the incorporation of research plots/stands that include desirable stand characteristics such as high BA spruce content as well as high stem density may provide a path towards gaining further insight into how to effectively and efficiently produce high quality tonewood via Sitka spruce.

It is also important to note that the sample sizes between the three GDRco locations were significantly different in size (see full data set in appendix) and further, there was outlier data that occurred at the R-Line, Eureka location, stem #26. While it was just outside of the methodology parameters for DBH, it was within the transect and sampled due to its canopy position. The RPIs for this sample were much greater than any other samples taken at R-Line Eureka and greater than any taken across all of the GDRco samples (Table 6). With this stem removed, the mean RPIs across all three locations are not significantly different from one another.

## ACF Quality

A regression model as well as a General Linear Model was constructed for this data set but neither resulted in significant outcomes. Results of the ACF sample responses to predictability were not significant or stronger than those originating from the GDRco sample group except age. Individual regressions were performed for all independent variables against the dependent RPI variable (Figure 7, Table 7).

The results of the ACF sample set added further validity that age is the strongest predictor to RPI response, however, none of the reported $R^2$ values suggest a strong model fit. While the results do indicate that age is likely the largest determining factor for this analysis, it is important to note that the sample size for this location was significantly smaller than those from GDRco locations. Additionally, samples at this location were only gathered in hopes of locating successful RPI minimums within the regional sample area and to aid in understanding RPI production with differing ownership objectives and silvicultural prescriptions.

*Figure 7. Regression analysis with ACF RPI as the response and Age as the predictor.*

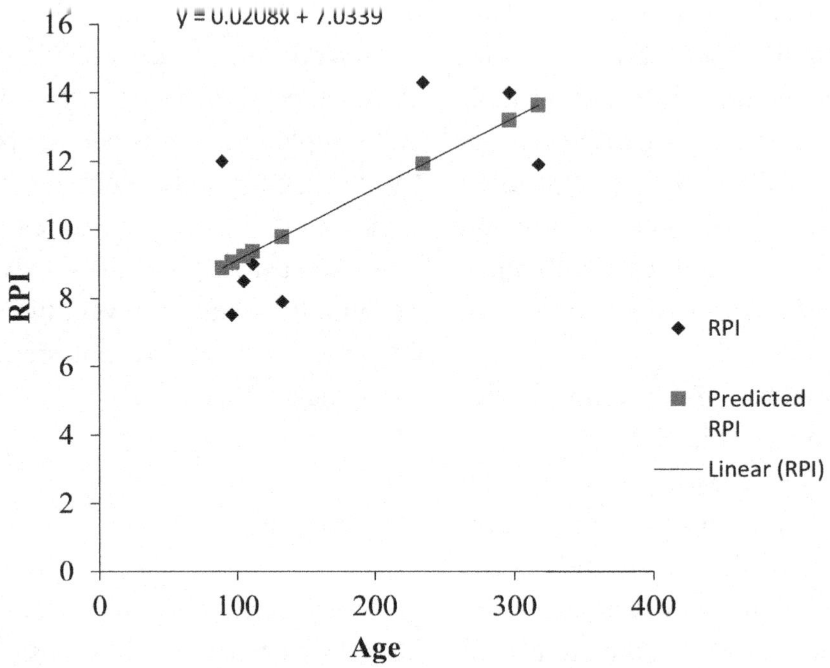


*Table 8. Individual regression analysis results for all independent ACF variables.*

| Source | *DF* | *SE* | *F-Value* | *P-Value* | $R^2$ adj (%) |
|---|---|---|---|---|---|
| ACF Age | 8 | 1.914 | 8.054 | 0.025 | 46.86 |
| ACF DBH | 8 | 2.119 | 5.277 | 0.055 | 34.84 |
| ACF Height | 8 | 2.649 | 1.517 | 0.264 | 6.88 |

The ACF is actively managed for multiple objectives and not just for forest improvement or monetary output. However, timber harvests have and do take place via a thinning or single tree selection method under an uneven-age silviculture system (ACF FMP, 2020). The general status of the forest as mature second growth (Iberle et al., 2020) signals that a longer rotation cycle is the key underlying element in achieving the desired RPIs. While the management objectives and silvicultural prescriptions are quite different between GDRco and the ACF, there is sufficient data to suggest that by extending rotation time and incorporating a mixed silvicultural system, GDRco may be able to produce high quality tonewood in the future.

## OP Quality

Olympic Peninsula (O)P data analysis produced a wide range of results but was ultimately not included in individual sample location analysis as it was beyond the scope and time constraints of this thesis. The purpose of collecting samples from the Olympic peninsula was simply to compare the mean RPI difference between old growth and second growth (both young and mature) trees.

## Quantity Analysis

Growth for acoustic guitar sales in a ten-year trailing period reported a weak overall unit demand. Despite weaking demand for individual units, overall wholesale value trended up and concluded 2022 with a positive change of 39.39%. The ten-year retail value reported a 42.11% positive change. Total domestic units shipped reported the lowest positive change with a total of 9.74 % (Table 8). Yearly total volume requirements for domestic demand of steel string acoustic guitars utilizing spruce as the tonewood for sound production as well as spruce for internal bracing totaled to 1.69 MMBF (Table 9).

Confidential internal inventory data supplied by GDRco was manipulated and analyzed. Outputs reported a pessimistic approximate yearly average output of 1.92 MMBF with 18" DBH or greater. Due to the specific and stringent processing requirements for producing spruce tonewood tops, it is important to note that the production of tonewood billets requires the logs to be quarter-sawn. This type of processing produces significant wastage and analyzing whole logs and assuming high utilization rates for tonewood production as was done with this analysis should be approached with caution. However, the remaining section of the log not quatersawn may be used in the production of internal bracing for acoustic guitars and by combining these two manufacturing needs, utilization would significantly improve. Due to time

*Table 9. 10 years of domestic acoustic guitar production and sales data. Source: Music Trades.*

| Year | Total Unit Volume (millions) | Spruce Unit Volume (millions) | Wholesale Value ($) | Retail Value ($) | Avg. Unit Price (retail, $) |
|---|---|---|---|---|---|
| 2022 | 1.51 | 1.13 | 666,816,000.00 | 1,041,900,000.00 | 690.0 |
| 2021 | 1.85 | 1.39 | 698,560,000.00 | 1,091,500,000.00 | 590.0 |
| 2020 | 1.60 | 1.20 | 641,456,000.00 | 957,398,000.00 | 598.0 |
| 2019 | 1.54 | 1.16 | 577,808,000.00 | 862,400,000.00 | 560.0 |
| 2018 | 1.49 | 1.12 | 549,065,000.00 | 819,500,000.00 | 550.0 |
| 2017 | 1.51 | 1.13 | 495,733,000.00 | 739,900,000.00 | 490.0 |
| 2016 | 1.39 | 1.04 | 456,337,000.00 | 681,100,000.00 | 490.0 |
| 2015 | 1.45 | 1.09 | 466,320,000.00 | 696,000,000.00 | 480.0 |
| 2014 | 1.50 | 1.12 | 454,506,000.00 | 678,368,000.00 | 453.0 |
| 2013 | 1.36 | 1.02 | 404,141,000.00 | 603,195,000.00 | 442.0 |
| Total avg. unit volume | 1.52 | 1.14 | Avg. yearly spruce units | | |
| **Percent change 10 yr.** | | | | | |
| Units Sold | 9.74 | | | | |
| Wholesale Value | 39.39 | | | | |
| Retail value | 42.11 | | | | |

*Table 10. Individual unit material requirements as well as total yearly dimensional lumber needs to meet domestic acoustic guitar demand.*

| Standard Tonewood Billet Size | | $CM^3$ | $FT^3$ | BF | Yearly BF Required | MMBF |
|---|---|---|---|---|---|---|
| Top (cm) | 54.5 x 22 x 2.1 | 0.003 | 0.089 | 1.068 | 1686417.90 | 1.686 |
| Brace (cm) | 47.4 x 6 x 1.9 | 0.001 | 0.019 | 0.229 | | |

constraints, determining the specific wastage and utilization for tonewood billet production was beyond the scope of this research, however, it is generally assumed that approximately 40 - 50% of a log is suitable for tonewood billet production (Gore, 2011).

The robust supply of Sitka spruce currently available across the norther coast region and specifically within the GDRco California ownership is significant and would be more than adequate in supplying a considerable volume to domestic acoustic guitar manufacturing needs. While the baselevel quantity does exist, there would need to be significantly more research and investments made into understanding how to best meet the quality needs. Examining the 10-years of sales data (Figure 8) and taking into account the unprecedented impact of Covid-19 on retail sales and supply, growth in the acoustic steel-string market should continue at a moderate pace for the foreseeable future and help to negate any concerns about long-term market variability. Access the manufacturing markets with local supplies of Sitka spruce may provide a welcomed change to a market that continually demands more attention to local resources (French et al., 2009).

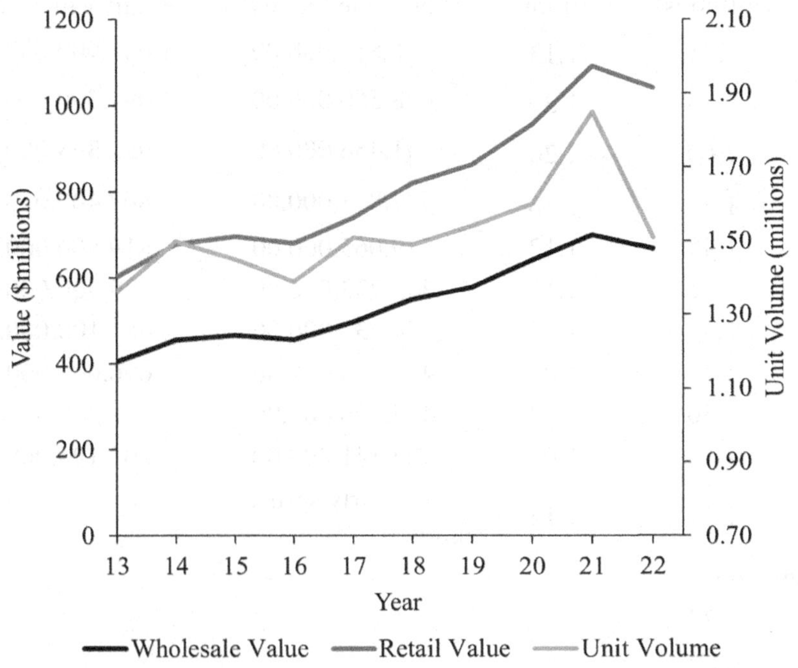


*Figure 8. 10-year change of wholesale value, retail value, and number of units shipped domestically per year.*

## Conclusion

Results of the full data analysis point to several key variables that have the most significant impact on producing stems with the RPIs and volume required for high quality tonewood tops (Table 2). The coastal regions of Humboldt county have the ability to produce fast growing trees. For traditional dimensional lumber markets growth of this magnitude is encouraged, but it is conflicting in the production of stems that produce high quality tonewood tops. Allowing stands, or nested stands within larger units to remain dense, limiting light exposure and increasing rotation age should all be considered if producing top wood is the main objective. Several of the units located in the Big Lagoon plots had considerable BA of dense, intermediate – suppressed canopy class spruce stems. The stems were far short of the minimum requirements presented in this methodology, but do represent a desirable starting place for further research. By forgoing at PCT and a CT, and utilizing only a single entry at final harvest, GDRco may be able to provide the necessary growing conditions for optimal RPI production. These types of stand conditions are highly advantageous for implementing additional research and gaining further understanding of the requirements for critical RPI and volume production on a commercial scale.

Another possibility exists for stands that do not have significant BA % of spruce. Spruce located within these stands should also be left through PCT and CT regardless of crown position within the stand. Applying both of these modifications to current silvicultural regimes across GDRco ownership may result in tremendous volumetric gains of stems suitable for tonewood production. There should be considerably more research put into how best to effectively address the long-term production and sustainability of these highly specific quality requirements as identifying sporadic individual stems that meet both the age and diameter minimums is not sustainable.

With age being the most significant indicator in RPI production (Figure 5), the shorter rotations currently employed by GDRco may not suit the requirements in developing top wood production as a dependable revenue stream in the short-term future. However, the brace wood, which constitutes approximately $1/5^{th}$ of the required spruce per instrument (Table 9), does provide GDRco with an option to create a product in the short-term future and allow for entry into the raw materials market for acoustic guitar production. RPI requirements for brace wood are much lower and fall within the mean RPIs from the GDRco sample locations (Table 5, Figure 6). Should forgoing a PCT and or a CT be unavoidable, the small diameter spruce harvested at those entries has the potential to be utilized for brace wood production.

Additional research is needed to gain further understanding into the specificity needed for brace wood production and how that may be best applied to commercial timberlands. It is recommended that a cost benefit analysis be performed to determine whether allocating a specific percentage of stands with high BA % of spruce may be transitioned in favor of longer rotations and converted to acoustic instrument billet production. It is unknown at this time whether that conversion would result in a positive gross expected revenue outcome.

The mature timberlands that GDRco manages are more than adequate to provide a significant percentage of raw materials required by the domestic steel-string acoustic manufacturing market. Their proximity to larger number of man-

ufacturers located within California and Oregon, including the current largest global producer of steel-string acoustic guitars, provides them with a unique position to potentially lower manufacturing acquisition prices by providing a product that is much closer to obtain. It may also help to drive positive public opinion towards continued development of more sustainable resource procurement by the guitar industry. Through sourcing material from this region, it may also allow GDRco to diversify their product offerings and add additional sources of revenue to their portfolio.

Several limitations exist within the body of this research and data set. Limited time for data acquisition resulted in a smaller than ideal data set for statistical analysis. A much larger sample set is recommended for continued research on this thesis. The limited data is evident in the R-Line, Eureka sample location, with only four stems sampled. It is import to note this location because stem #26 was an outlier to the data. Removing this stem significantly reduces the mean RPI for the sample location and the new results of mean RPIs for GDRco locations were insignificant. In addition, amendments to the methodology should be made to capture smaller DBH stems and those occupying the suppressed – intermediate crown classes. Efforts should also be made to destructively sample stems cored that present encouraging RPI numbers and convert them into logs and billet wood for further analysis. Density testing as well as modulus of elasticity testing should be conducted on the logs to establish true soundwave propagation potential.

*Table 11. Full data set for all samples taken during in the field and those provided by CPH.*

| Sample Location | Total Stem # | Unit Stem # | DBH (in) | Total Height (ft) | Clearwood Height (ft) | Crown Ratio % | Crown Class | Mean RPI | Age (yr) |
|---|---|---|---|---|---|---|---|---|---|
| Big lagoon | 1 | 1 | 37.3 | 154.2 | 0 | 90 | D | 4 | 38 |
| Big lagoon | 2 | 2 | 34.2 | 139.6 | 0 | 90 | D | 3 | 33 |
| Big lagoon | 3 | 1 | 39.9 | 149.7 | 42 | 60 | D | 2 | 46 |
| Big lagoon | 4 | 2 | 39.7 | 147.2 | 63 | 50 | D | 2.5 | 35 |
| Big lagoon | 5 | 3 | 29.1 | 164 | 35 | 70 | CD | 4 | 41 |
| Big lagoon | 6 | 4 | 28.5 | 158.3 | 32 | 60 | CD | 5 | 42 |
| Big lagoon | 7 | 1 | 31.4 | 171.1 | 0 | 90 | CD | 3 | 41 |
| Big lagoon | 8 | 2 | 33.8 | 159.3 | 31 | 70 | CD | 4 | 39 |
| Big lagoon | 9 | 3 | 27.9 | 155.6 | 12 | 90 | CD | 2 | 38 |
| Big lagoon | 10 | 4 | 34.3 | 149.3 | 16 | 90 | CD | 4.5 | 39 |
| Big lagoon | 11 | 5 | 25.7 | 151.9 | 10 | 80 | CD | 3.5 | 41 |
| Big lagoon | 12 | 1 | 24.6 | 106.8 | 25 | 60 | CD | 4 | 39 |
| Big lagoon | 13 | 2 | 23.9 | 119.5 | 41 | 40 | CD | 3 | 40 |
| Big lagoon | 14 | 3 | 21.4 | 112.7 | 26 | 40 | CD | 4 | 37 |

| Sample Location | Total Stem # | Unit Stem # | DBH (in) | Total Height (ft) | Clearwood Height (ft) | Crown Ratio % | Crown Class | Mean RPI | Age (yr) |
|---|---|---|---|---|---|---|---|---|---|
| Big lagoon | 15 | 4 | 31.7 | 131.1 | 21 | 60 | D | 3 | 49 |
| Murray | 16 | 1 | 23.3 | 162.1 | 12 | 80 | CD | 4 | 45 |
| Murray | 17 | 2 | 31 | 161.4 | 16 | 80 | D | 4.5 | 42 |
| Murray | 18 | 3 | 35.3 | 138.9 | 32 | 60 | D | 4 | 36 |
| Murray | 19 | 4 | 19.9 | 128.1 | 28 | 50 | CD | 3.5 | 40 |
| Murray | 20 | 5 | 25.5 | 162.4 | 52 | 30 | CD | 3 | 35 |
| Murray | 21 | 6 | 19 | 99 | 20 | 50 | CD | 2.5 | 32 |
| Murray | 22 | 1 | 28.6 | 167.2 | 28 | 60 | CD | 3.5 | 39 |
| Murray | 23 | 2 | 18.1 | 144.1 | 8 | 90 | I | 5.5 | 37 |
| Murray | 24 | 3 | 28.3 | 152.5 | 8 | 90 | CD | 2 | 36 |
| Eureka | 25 | 1 | 22.3 | 116.4 | 60 | 40 | CD | 4 | 37 |
| Eureka | 26 | 2 | 18.9 | 75.1 | 32 | 20 | I | 8 | 27 |
| Eureka | 27 | 3 | 26.6 | 118.3 | 44 | 40 | CD | 5 | 44 |
| Eureka | 28 | 4 | 21.9 | 117.3 | 35 | 50 | CD | 5.5 | 43 |
| ACF | 29 | 1 | 51.3 | 179.4 | 89 | 30 | CD | 9 | 112 |
| ACF | 30 | 2 | 29.2 | 174.3 | 58 | 50 | CD | 8.5 | 105 |
| ACF | 31 | 3 | 22.9 | 144.8 | 72 | 40 | CD | 9 | 97 |
| ACF | 32 | 4 | 22 | 128.8 | 74 | 40 | I | 12 | 89 |
| CPH - ACF | 33 | 1 | 130.7 | 284.1 | | | | 11.9 | 318 |
| CPH - ACF | 34 | 2 | 35.7 | 163.6 | | | | 7.5 | 96 |
| CPH - ACF | 35 | 3 | 90 | 190.9 | | | | 14.3 | 235 |
| CPH - ACF | 36 | 4 | 119.2 | 307.4 | | | | 14 | 297 |
| CPH - ACF | 37 | 5 | 58.2 | 214.9 | | | | 7.9 | 133 |
| CPH - OP | 38 | 1 | 87.1 | 264.7 | | | | 11.8 | 256 |
| CPH - OP | 39 | 2 | 88.5 | 266.2 | | | | 13.7 | 288 |
| CPH - OP | 40 | 3 | 125.4 | 295.7 | | | | 15 | 353 |
| CPH - OP | 41 | 4 | 75.3 | 287.4 | | | | 15.5 | 321 |
| CPH - OP | 42 | 5 | 116.6 | 257.9 | | | | 8.7 | 317 |

## References


Youngs, Robert L. 2003. Wood Science and Technology in North America. *Forest Products Journal* 53, (11) (Nov): 12-21,

Gibson, Chris, and Andrew Warren. "Resource-sensitive global production networks: reconfigured geographies of timber and acoustic guitar manufacturing." *Economic Geography* 92, no. 4 (2016): 430-454.

Taylor Guitars. "Taylor Guitars: The Ebony Project in Cameroon." The Ebony Project, 2018. https://www.taylorguitars.com/ebonyproject/.

Music Trades. "A Conversation with Martin Guitar CEO Thomas Ripsam, Sustainability Issues." Edited by Brian Majeski . Youtube, February 27, 2024. https://www.youtube.com/watch?v=akpjNjo1UlU.

Bennett, Bradley C. "The sound of trees: wood selection in guitars and other chordophones." *Economic Botany* 70 (2016): 49-63.

Schloss, Joseph G, Larry Starr, and Christopher Waterman. "5." Essay. In *Rock: Music, Culture, and Business*, 124–36. New York, New York: Oxford University Press, 2012.

French, Mark, Rod Handy, and Mark J. Jackson. "Manufacturing sustainability and life cycle management in the production of acoustic guitars." *International Journal of Computational Materials Science and Surface Engineering* 2, no. 1-2 (2009): 41-53.

Badiane, David Giuseppe, Sebastian Gonzalez, Raffaele Malvermi, Fabio Antonacci, and Augusto Sarti. "A neural network-based method for spruce tonewood characterization." *The Journal of the Acoustical Society of America* 154, no. 2 (2023): 730-738.

Stephens, Howard Page. "The Effect of Finishes on the Vibration Properties of Spruce Guitar Soundboard Wood." *Savart Journal* 1, no. 5 (2015).

Kevin Ryan. Ryan Guitars. Master Luthier, Owner of Ryan Guitars. Ryan Cooper. Email Correspondences. February, 2024.

Drew Haines. Former Head Luthier Guild Guitars. Ryan Cooper. Email Correspondences. January – February, 2024.

Haynes, Richard W. *An analysis of the timber situation in the United States: 1952 to 2050*. Vol. 560. US Department of Agriculture, Forest Service, Pacific Northwest Research Station, (2003).

FASM, James D. Cotton, and John D. Wolodko FASM. "Identification lf Sustainable Tonewoods for Acoustic Guitars Using Materials Selection Software." *Advanced Materials & Processes* 181, no. 5 (2023): 13-18.

Calvano, Silvana, Francesco Negro, Flavio Ruffinatto, Daniel Zanuttini-Frank, and Roberto Zanuttini. "Use and sustainability of wood in acoustic guitars: An overview based on the global market." *Heliyon* (2023).

Pedgley, Owain, and Eddie Norman. "Materials innovation in acoustic guitars: challenging the tonal superiority of wood." *Leonardo Music Journal* 22 (2012): 1-24.

Music Trades. "The Music Industry Census: Dollar and Unit Volume Data On Top Product Categories", 2023. Distributed by The Music Trades Corporation.

Garzon Arbelaez, Melina. "Rosewood trade: A case study of the imports into the United States." PhD diss., 2022.

Innes, John L. "Madagascar rosewood, illegal

logging and the tropical timber trade." *Madagascar Conservation & Development* 5, no. 1 (2010).

Merchel, Sebastian, M. Ercan Altinsoy, and David Olson. "Perceptual evaluation of bracewood and soundboard wood variations on the preference of a steel-string acoustic guitar." *The Journal of the Acoustical Society of America* 146, no. 4 (2019): 2608-2618.

Martinez-Reyes, Jose. "Mahogany intertwined: Enviromateriality between Mexico, Fiji, and the Gibson Les Paul." *Journal of Material Culture* 20, no. 3 (2015): 313-329.

Gore, Trevor. "Wood for guitars." In *Proceedings of Meetings on Acoustics*, vol. 12, no. 1. AIP Publishing, 2011.

Brackley, Allen M, and Richard W Haynes. "Timber Products Output and Timber Harvests in Alaska." Government Info, August 2008. https://www.govinfo.gov/content/pkg/GOVPUB-A13-PURL-LPS113505/pdf/GOVPUB-A13-PURL-LPS113505.pdf.

## Additional Supplementary Resources

Dunn, Chris. "Analyzing the Acoustical Properties of Alternative Materials in Guitar Soundboards to Reduce Deforestation." Thesis, California Polytechnic State University, San Luis Obispo, 2013.

Margetts, Rebecca, and Michael James. "Quantifying the Properties of Guitar Tonewoods." Quantifying the Properties of Guitar Tonewoods, May 13, 2023. https://www.aes.org/e-lib/browse.cfm?elib=22101.

Cargill , Nicholas. Assessing the Efficacy of CITES Restrictions on Malagasy Rosewood , April 29, 2022. https://scholarship.richmond.edu/cgi/viewcontent.cgi?article=2688&context=honors-theses.

Yoshikawa, Shigeru. "Acoustical Classification of Woods for String Instruments." *The Journal of the Acoustical Society of America* 122, no. 1 (July 1, 2007): 568–73. https://doi.org/10.1121/1.2743162.

Trump, Maxine, Josh Granger. *Musicwood.* Helpman Productions, 2013. https://musicwoodthefilm.com/.

IdeaFest
Journal

# The Online Education Landscape, Digital Tools, and COVID-19

**Rosetta Burk**


## Abstract

The 2020 COVID-19 pandemic has been the ultimate stress test for communication technologies, online learning methods, digital pedagogies, students and educators as the world faced school closures and emergency online learning situations during the pandemic lockdowns. There has been an influx in communication tools, digital paradigms, and teaching methodologies in addition to relying on existing digital tools and methods to continue students' education while unable to meet in person. The level of remote learning and digital movement is unprecedented and the results have been varied and illuminating. Research has been conducted on student attitudes in the immediate aftermath of the emergency transition, on specific teaching tools, and on remote instruction in regards to learning outcomes, but the long term effects of the unprecedented circumstances of COVID-19 lockdowns on student's educational development and prospects are still unknown. There are also the existing privacy and safety concerns for digital tool interaction that extend to education that have been exacerbated by the emergency transition. This paper will explore and evaluate the current state of online learning and the digital tools that facilitate it, as well as the recent developments in remote learning due to the emergency transition to online learning. It also aims to evaluate the effects on a student's quality of education and wellbeing.

The year of 2020 began with a plague. In the Hubei province of China, cases of a "mysterious viral pneumonia" were reported in the city of Wuhan. Three months later the newly discovered coronavirus had a global death toll at 100,000 in dozens of countries and territories (BBC, 2020). This unfortunate local phenomena rippled into a global epidemic and the unforeseen consequences of the pandemic rippled out with it. Lockdowns, mask mandates, travel restrictions, and travel advisories went into place swiftly. In the U.S and other states, work and school went remote. For teachers, educators, parents, and students, a new curriculum that could be carried out remotely had to be developed or adapted overnight. While possibly necessary, the emergency transition from in person to remote learning was not smooth; one day students were hearing about a "new flu" on the opposite hemisphere, the next, the rest of the school year was canceled because of an imminent pandemic. For many, including the author of this paper who was a student at the time, spring of 2020 was a lost semester, cut short by clumsily and quickly imposed safety measures.

Starting around March of 2020, 77% of public schools and 73% of private schools reported moving all or some classes to a remote learning model. Changes in schooling had some predictable patterns; urban schools reported the most movement to distance modes of learning and rural schools the least. Parents of younger students seemingly were less satisfied with the transition than those of older students. During the 2020-21 school year, public school 9th through 12th grade enrollment rates seemed unaffected but enrollment for pre-K through 8th grade dropped in every state with an average of 4%, more than a million students. There was a similar but converse uptick in homeschooling in the pre-K through 8th grade group (National Center for Education Statistics, 2020-21).

Despite 2021 having more cases of COVID-19 than 2020 and being the deadliest year in United States history because of COVID-19 related deaths, in part because of COVID-19 variants and vaccine hesitancy, schools still returned to in person modes of teaching (Stobbe, 2022). By May of 2021 half of all students were in full-time in-person schooling despite remote learning being offered in nearly 80% of cases (National Center for Education Statistics). While there are a variety of reasons why students, parents, and teachers prefer in-person enrollment, one could be how completely unprepared anyone was to move education online enmasse in the first place. Teachers were forced to make quick and important decisions regarding their curriculum with little to no training, precedent, or support. Instructors chose a variety of methods, formats, tools, and technologies to deal with the emergency transition and some strategies were more effective than others. Despite this, the circumstances offer an incredible insight into online teaching and digital communication tools, which seems to be here to stay and is proving to play a significant part in the future of education.

The most obvious insight for success in online teaching seems to be that quality of education is directly correlated with quality of communication. This communication is shaped by and facilitated by digital tools though one of the main barriers to online learning during the emergency transition was access to these tools. Many students did not have the technology, consistent internet access, or dedicated spaces in their homes to make full use of the tools being used. Regardless, according to early reporting the most common approach to adapting to the emergency transition from instructors was

to integrate existing lesson plans and assignments into a Learning Management System (LMS) while holding synchronous meetings (Gillis and Krull, 2020). A LMS is a software application or web-tool that is used to facilitate the learning process by hosting content, communication channels, and collecting data for assessment. Some common LMS' in the United States are *Canvas, Schoology,* and *Google Classroom*, though there are dozens of other applications and programs. LMS' have proven useful and were already being integrated into schools nationwide. A digital element like LMS' are a standard feature in schools. LMS' were used alongside synchronous meetings facilitated by video conferencing software like *Google Meet, Microsoft Teams,* or *Skype* but overwhelmingly by the application *Zoom*. Depending on the method used these scheduled meetings supplemented by Learning Management Systems can be extremely similar to an in-person course. Courses can also be asynchronous, relying on students to self pace through material using only digital tools and platforms.

Besides LMS' and conferencing applications they're are a variety of other tools instructors, students, and administrators use. School assigned emails are ubiquitous. There is commonly implemented third party online material: educational courses like "Khan Academy", educational sources like "Crash Course" youtube videos, or learning applications like *DuoLingo*. There are also countless study and note applications like *EverNote* or *Notability* that students and educators use. Accounts and access to data and file keeping software like *Google Drive, OneDrive,* or *Dropbox* are often offered through schools. Schools are spoiled for choice when it comes to online tools and materials

Many of these softwares, like *Microsoft* or *Google's* suite of collaboration and data management tools, have been commonplace in the education and business sector for years now. The COVID-19 pandemic has pushed this digitization into ubiquity, especially for schools, and this newfound dependence is intensifying concerns about the safety of digital tools. A tool might be a neutral force but the corporations, developers, shareholders, and businesses behind these tools, and more importantly the data they collect, are not.

A massive amount of data is collected by applications and software designed for education. For example, *Instructure's Canvas* is the most popular Learning Management System in the U.S for higher education and the self reportedly fastest growing for k-12. The 2022 Privacy Evaluation for Canvas by Common Sense Media reports that it automatically collects, amongst other data, personally identifiable information(PII), geolocation data, and usage analytics data; it's also unclear whether biometric or sensitive data is collected. *Instructure's Canvas* scored the worst in the data privacy area of the evaluation because personal information is sold or rented to third parties and it's unclear how this information is protected or what it's being used for. The ease of which student, teacher, and district data is shared from *Canvas* to and among third parties is facilitated by the LMS' "frictionless" open-source model. The "frictionless" user experience is based on the interoperability of systems from third-parties that can be integrated straight into the Canvas LMS from Canvas' Open Apps, a library of hundreds of third party applications that harvest and share data (Marachi and Quill, 2020). These are content apps like "Khan Academy", video conferencing apps like *Zoom*, and note taking apps like *Evernote*.

With the push from COVID-19 moving LMS' and other digital tools into k-12 and the corporate sphere, there is an ever expanding net for students that is dictating more and more of their lives. *Canvas* uses Amazon Web Services (AWS) to store and host all its cloud data and AWS has a program called *EdStart* that partners with hundreds of companies all over the world that create educational software products and also provides them with cloud data storage services. This creates a massive pool of cloud data pulled from hundreds of online sources that can create accurate and identifiable profiles of students. These LMS' are being integrated into every level of schooling. Independent tools that students use to supplement learning, take notes, and complete assignments are also collecting data. *Portfolium* is a tool from *Instructure*, the parent company of Canvas, used in thousands of colleges to connect students directly to employers by analyzing and allowing direct access to their academic history and previous work for career opportunities. *Bridge* is another Learning Management System from *Instructure* but this is a corporate LMS for employee training. It is conceivable that a student will have a constantly updated, uninterrupted, comprehensive profile of personally identifiable information, location, interests, academic and behavioral assessment, achievement, and ability from before elementary school to the time they enter the workforce. This is a new type of surveillance that students have been subjected to that's been exacerbated by the COVID-19 pandemic accelerating digitization of education; "dataveillance".

Educational digital tools are not the only way information is harvested from students. IPP and other data is gathered on individuals from social media interactions, online and real life purchases with credit cards, internet searches, GPS data from mobile phones, biometrics from CCTV or "smart" watches, and academic participation with education technologies to create accurate, actionable, and identifiable information on individuals and this data does not just sit in servers. This information is fed through algorithms to create "predictive analytics". These are machine inferences about a person's demographics, personal information, health, financials, or interests based on this conglomerated data (Crawford and Schultz, 2014). A now famous example is the chain store Target creating an algorithm with the statistician Andrew Pole that examines purchases and other accrued personal information in order to accurately predict pregnancies in customers. This piece of intimate information is then used to create a targeted and personalized advertising campaign (Duhigg, 2012). Accurate data regarding intimate or personal matters are generated not through normal modes of surveillance but through applying algorithmic analytics to mined data, hence "dataveillance". There are very real "predictive privacy harms" that come from dataveillance. The same way *Target* can identify groups of desirable customers and send personalized advertisements, credit companies and real estate agencies can identify groups of *undesirable* customers and withhold advertising. A person's opportunities in life are increasingly being shaped by machine inferences and predictions that gatekeep access to, amongst other things, employment, insurance, and credit (Crawford and Schultz, 2014). There are currently no protections from or avenues to redress predictive privacy harms but dataveillance is a present and growing issue for everyone and students are no exception.

Learning analytics are a prominent form of

dataveillance in education. Digital education tools harvest data from student interactions and academic progress and generate predictions for which students need intervention as well as predict a student's future progress (Siemens, 2013). *Instructure* promises it's product-

> "Elevate K-12 Analytics is the analytic application that combines data from key district data sources and provides access to highly interactive visualizations of integrated data so educators can measure, manage, and improve student outcomes."
> -*Elevate analytics*: K-12. Instructure. (n.d.).

As an aside, a student's academic performance is not the only metric by which students are measured and managed. *ClassDojo* is a gamified application and communication platform with tens of millions of students worldwide that is used to track and monitor student behavior in the classroom; incentivizing behaviors by giving and taking away "Dojo points", a digital profile of a student's behavior is created that is seamlessly shared between the school and home. There are also technologies that create biometric data either through gamified self-reporting or through wearable meters employed by schools. New methods are integrated with more classic forms of surveillance; facial recognition software being added to the ubiquitous CCTV cameras in schools and mobile phone tracking of a student's digital presence (Nemorin, 2017). Physically, emotionally, and cognitively, students are being surveilled.

In academic dataveillance, learning analytics platforms create real-time assessments that don't just predict a student's performance, many have automated recommendation systems to prescribe educational pathways, like Pearson and Knewton's "Adaptive Learning Platform". While injecting algorithmic learning and data-driven approaches into education is often justified by the "personalization" it provides students, it bears noting that this is not a real solution to a lack of individualized attention that students need to thrive and in fact homogenizes the student by reducing them to algorithmic assemblages. Digital quantification is a reductive process; when students are sorted, compared, and assessed by machine learning, entire dimensions of their being is lost or diminished. What is important is only what is in the eye of the software, the things that cannot be measured are not just devalued, they are invisibilized. The ability of learning analytics to project probabilities and possibilities can easily, and has been observed to, become prescriptive rather than descriptive. Students usually have little idea what data and how much data is being collected on them and so they cannot dispute the claims, projections, or profiling of algorithms and the potentialities of students are flattened by the social sorting inevitable with dataveillance (Lupton and Williamson).

Despite the image painted so far of the online learning landscape, this paper does not intend to fearmonger about "Big Data" replacing teachers or children being shuffled through life by artificial intelligence in a digital panopticon. All the tools described here, despite possibly multitasking as modes of surveillance, are also generally just tools being used to serve a pedagogy designed by human educators. The gamification of learning is not particularly new and students can get genuine enjoyment and benefit from these gamified learning applications (Whitson, 2013). Additionally, tracking and datifying students isn't new either, governments the world over have been tracking student progress for centuries through tests, evaluations, and surveys to find patterns and address issues in order to create proficient citizens; Learning Analytics and Educational Data Mining simply make procuring and examining this data much easier and more comprehensive. These are still burgeoning fields of study and can be used to identify patterns on the micro, meso, and macro level in education and improve learning outcomes by identifying and suggesting data driven solutions. When interpreted by educators trained in this kind of data analysis this can be a beneficial thing (Sghir et al., 2022). Attention to the potential pitfalls and harms of these tools are coming into prominence at the same time as the tools themselves and there are also measurable benefits to online learning and the digital tools that facilitate it.

There are some unique benefits to remote learning, the most obvious being its titular quality of being remote. The flexibility and freedom allowed by schooling no longer being tied to a geographic location, or particular times in the case of asynchronous classes, is one of the greatest draws of online school. It broadens what education can offer and therefore serves students who are under-resourced and under-served the most. Rural students get the chance for no or shorter commutes and a broader range of classes that leads to greater opportunity in the future (Beck et al., 2018). Online schooling is especially prolific in higher education because the convenience and economy of online school serves adults who are working or have other responsibilities. There are also the students who simply prefer the online modality. Many students and parents find the comfort, freedom, and flexibility of going to school from home more than worth the drawbacks. Having more choices is a good thing; remote learning fills in gaps in education for non-traditional students and has proven to be an effective method of learning on par with more traditional methods. Quality of learning outcomes comes down to the quality of education not the modality of the education (Larson et al., 2023).

While all learning is an evolving practice and a constant experiment to see what will most benefit endlessly diverse students, online learning is still being studied and developing with digital technologies and tools to see what the modality itself can offer outside of simply emulating traditional schooling. A unique and promising aspect in digital pedagogy not generally afforded in traditional schooling is the opportunity for anonymity allowed by digital tools. Active learning is considered the preferred method of learning to cultivate higher thinking skills but it requires real-time discussion with others. The traditional method is class discussions but many if not most students have apprehension over sharing in class for fear of being judged. Fear of Negative Evaluation (FNE) is the spectrum of anxiety students feel when they are uncertain of how they will be evaluated by peers and teachers and it's cited as the top reason why students don't participate in class. This keeps them from receiving the benefits of discussion and argumentation that develops these critical skills. FNE is combated by creating psychological security which partial or total anonymity can provide. Digital education games like *Kahoot!* that allow student's answers to remain anonymous promote engagement and ease anxiety. Where a voice is required for discussion, applications like *Clubhouse* and camera-off *Zoom* rooms can facilitate these critical discussions while also fostering psychological security through partial anonymity (Yep et al., 2023).

Online learning and digital tools have unique advantages as a modality but this was not reflected in the emergency transition to remote learning during the 2020 pandemic. Calculating the costs of this emergency transition in retrospect, many consider it something of a failure and increasingly, many are wondering if it was even necessary. The historical precedent for closing down schools in times of mass sickness comes from studies and policies generated during Influenza pandemics. Children are much more susceptible to Influenza making these policies very effective, however, children were and are less likely to contract the SARS-CoV-2 virus in comparison to other groups. So much so that taking into consideration other measures to reduce interactions taken at the time and during school reopenings, there is evidence that the mass school closures were not necessary (Wu et al., 2021). There were also other things to consider like the massive and unprecedented strain on healthcare workers who were suddenly also saddled with looking after their children in a time where there was almost no childcare available (Bayham & Fenichel, 2020).

While there is some contention over whether school closures were effective in stopping or slowing the spread of COVID-19, there is little dispute on the educational damages and delays caused by mass school closures.There was rampant learning loss during this time as students and teachers were forced into remote learning situations on the fly. Damage was done quickly and recovery has been slow. After school reopenings there was an increase in chronic absenteeism (Dee, 2024) and dropouts (Moscoviz & Evans, 2022). Learning harms were not distributed equally either; first-generation and minority students fared worse during this time, facing more barriers to online learning like lack of internet or a dedicated study space as well suffered worse under the collective dip in mental health, increasing the learning inequality gap. While the profound mental health ramifications students were dealing with at the time weren't distributed equally, it seems that all students were feeling unmotivated during such uncertain and anxious times (Gillis and Krull, 2020).

While the emergency transition to remote schooling in 2020 seems to be something of a failed experiment because of the extenuating circumstances, online schooling as a modality and digital education tools are here to stay. There are a vast array of digital tools and educational technologies available for students and educators but it is up to the instructor and student's dedication to communication to make these tools useful. The tools themselves are a work in progress in regards to their accessibility and long term ramifications on privacy and safety. Education is just one of the spheres affected by dataveillance and predictive analytics in the form of learning analytics and educational data mining and the predictive privacy harms and how to navigate these vast data mining complexes is something that will have to be addressed as we steam ahead in the digital age. The digital age offers educational opportunities as well, not just in the convenience or economy of being able to go to school wherever there is internet connection but a new educational modality with unique features that can improve learning in some areas and broadens who is capable of receiving an education and how.

Despite the estimated learning losses, calculated in dropped GPAs and testing points, and the new contention over whether school closures were necessary, there is no quantification for the value the lives and health of students and their families that was potentially preserved by emergency school closures. There is also no way of extricating the psychological effects of the underlying situation from the psychological effects of an unexpected move from traditional to online schooling. School is a critical space for students, only second to the home, where social, psychological, emotional, and often physical and nutritional needs are met; it is no surprise that being forced from this space causes adverse mental health effects. At the same time, students were bearing witness to turmoil, death, sickness, and political upheaval on the global stage, and that's just those who did not experience it firsthand. It is impossible to know the true and entire breadth of consequences from this global educational experience, good and bad. All that's left is to take this as a learning experience as we attempt to navigate the new digital age.

## References

Bayham, J., & Fenichel, E. P. (2020). Impact of school closures for covid-19 on the US health-care workforce and net mortality: A modelling study. *The Lancet Public Health, 5*(5). https://doi.org/10.1016/s2468-2667(20)30082-7

BBC. (2020, September 29). Covid-19: Milestones of the global pandemic. *BBC News*. https://www.bbc.com/news/world-54337098

Beck, D., Maranto, R., & Shakeel, M. D. (2018). Does rural differ? comparing parent and student reasons for choosing Cyber Schooling. *The Rural Educator, 37*(3). https://doi.org/10.35608/ruraled.v37i3.243

Common sense privacy evaluation for Canvas. The Common Sense Privacy Program. (n.d.). https://privacy.commonsense.org/evaluation/canvas

Dee, T. S. (2024). Higher chronic absenteeism threatens academic recovery from the COVID-19 pandemic. *Proceedings of the National Academy of Sciences, 121*(3). https://doi.org/10.1073/pnas.2312249121

Duhigg, C. (2012, February 16). How companies learn your secrets. *The New York Times*. https://www.nytimes.com/2012/02/19/magazine/shopping-habits.html

Elevate analytics: K-12. Instructure. (n.d.). https://www.instructure.com/k12/products/elevate/elevate-k12-analytics

Larson, M., Davies, R., Steadman, A., & Cheng, W. M. (2023). Student's Choice: In-person, online, or on demand? A comparison of instructional modality preference and effectiveness. *Education Sciences, 13*(9), 877. https://doi.org/10.3390/educsci13090877

Lupton, D., & Williamson, B. (2017). The datafied child: The dataveillance of children and implications for their rights. *New Media and Society, 19*(5), 780–794. https://doi.org/10.1177/1461444816686328

Marachi, R., & Quill, L. (2020). The case of canvas: Longitudinal datafication through Learning Management Systems. *Teaching in Higher Education, 25*(4), 418–434. https://doi.org/10.1080/13562517.2020.1739641

Moscoviz, L., & Evans, D. (2022). Learning Loss and Student Dropouts during the COVID-19 Pandemic: A Review of the Evidence Two Years after Schools Shut Down. Center for Global Development. https://www.cgdev.org/publication/learning-loss-and-student-dropouts-during-covid-19-pandemic-review-evidence-two-years

Nemorin, S. (2017). Post-panoptic pedagogies: The changing nature of school surveillance in the Digital age. *Surveillance and Society, 15*(2), 239–253. https://doi.org/10.24908/ss.v15i2.6033

Sghir, N., Adadi, A., & Lahmer, M. (2022). Recent advances in Predictive learning analytics: A decade systematic review (2012–2022). *Education and Information Technologies, 28*(7), 8299–8333. https://doi.org/10.1007/s10639-022-11536-0

Siemens, G. (2013). Learning analytics. *American Behavioral Scientist, 57*(10), 1380–1400. https://doi.org/10.1177/0002764213498851

Stobbe, M. (2022, April 12). *Covid helped make 2021 the deadliest year in U.S. history*. PBS. https://www.pbs.org/newshour/nation/covid-helped-make-2021-the-deadliest-year-in-u-s-history

Whitson, J. R. (2013). Gaming the quantified self. *Surveillance & Society, 11*(1/2), 163–176. https://doi.org/10.24908/ss.v11i1/2.4454

Wu, J. T., Mei, S., Luo, S., Leung, K., Liu, D., Lv, Q., Liu, J., Li, Y., Prem, K., Jit, M.,Weng, J., Feng, T., Zheng, X., & Leung, G. M. (2021). A global assessment of the impact of school closure in reducing COVID-19 spread. *Philosophical Transactions of the Royal Society A: Mathematical, Physical and Engineering Sciences, 380*(2214). https://doi.org/10.1098/rsta.2021.0124

Yep, B. L., Tan, T. K., & Fung, F. M. (2023). How partial anonymity may reduce students' anxiety during remote active learning—a case study using clubhouse. *Journal of Chemical Education, 100*(2), 459–468. https://doi.org/10.1021/acs.jchemed.2c00051

## National Association of Communication: Learning Outcomes (LOC)

- LOC #4 : Create messages appropriate to the audience, purpose, and context
- LOC #7: Apply ethical communication principles and practices
- LOC #9: Influence public discourse

*Learning outcomes in communication*. National Communication Association. (2022, March 9). https://www.natcom.org/learning-outcomes-communication. A series of PDF resources from the NCA's Learning Outcomes in Communication Project

IdeaFest
Journal

# Influence of Elevation on Plant Survivorship in Baduwa't Estuary, CA

**Celine Aguilera, Claire Durbin, Rick Toledo, and Rivé Prater**

## Abstract

The Baduwa't is a river in Humboldt County, California that has been adversely impacted by agricultural operations. Disturbances from agriculture have altered the surrounding ecosystems and hydrology of the river. The conservation stewardship group, California Trout, initiated a collaborative restoration project that created a new off-channel pond to provide habitat for salmonids and other native fish in the estuary. As an extension of this project, sections of the site underwent revegetation efforts. The survivorship rates of this new vegetation were lower than anticipated. One potential theory for low survivability was thought to be elevation and access to water. This research identified if the elevation on the site could be influencing plant success rates. To determine if proximity to groundwater through elevation was the cause, we assessed plant statuses and compared them to an elevation map utilizing ArcGIS Pro. After comparing the results, we did not find any clear correlation between elevation and plant success. We then compiled a list of potential recommendations such as mulching, the use of mycorrhizae, the influence of canopy cover, shelterwood, and various planting methods.

## Introduction

The Baduwa't is a river in Humboldt County, California that was known as the "Mad River" by settlers following colonization of the region. We will be strictly referring to it as Baduwa't because we believe it is important to use the name that was given to it by the Indigenous peoples' and first human inhabitors of this place. The indigenous peoples of this area have always had a strong connection with the Baduwa't. Through many generations, it has been a place of cultural and spiritual significance for these stewards of the area who directly helped the place to thrive with biodiversity. In contrast, the name "Mad River" comes from Dr. Josiah Gregg who had lost his temper because his exploration party did not wait for him at the river's mouth. This name stuck but has no significance to anything except a settler who was having a bad day, so we believe that it is crucial to refer to it by its correct Indigenous name, Baduwa't.

*Photo of Baduwa't and restored off-channel pond, taken by Celine Aguilera, March 17, 2024*

Estuaries are where the river meets the ocean, creating brackish water or a mixture of saltwater and freshwater (Basic Information, n.d.). They are a crucial ecosystem, containing nursery habitats for diverse fish species and hotspots for human activity since they provide many ecosystem services (Hodgson et al., 2020). Additionally, estuaries are biologically productive, so they are crucial places for migratory birds to stop and recharge for the rest of their journey (Basic Information, n.d.). They serve as natural barriers between the land and sea because of their ability to ease flooding (Basic Information, n.d.). They are also great nurseries since they provide a protective buffer from large ocean waves and storm surges. Specifically, juvenile salmon use estuaries to recharge and escape from predators before making their way to the ocean (Hodgson et al., 2020). In the last century, global estuaries have lost over 90% of species and native vegetation has been reduced to less than half of its historical size, stemming from habitat loss (Hodgson et al., 2020). This habitat loss can be traced to extractive and disconnected human activity. Historically, the lower Baduwa't had access to a broad floodplain that provided a critical rearing habitat for juvenile fish, supported a diverse array of wildlife, and was a vital food source for the Indigenous peoples of the land. Under Indigenous stewardship, estuaries such as Baduwa't thrived and were diverse, productive environments crucial to their way of life. Unfortunately, post-European colonization, estuaries like Baduwa't are some of the most damaged ecosystems and need both restoration and a return to relationships with Traditional Ecological Knowledge.

Influenced by natural and anthropogenic events, the Baduwa't watershed has been changed and altered since human habitation (Stillwater

Sciences, 2010). During pre-colonial times, the climate experienced warmer temperatures and drier vegetable regimes which dominated this period as we see today (Stillwater Sciences, 2010). The climate at that time was similar to our current conditions but supported larger stand conifers with higher rainfall and cooler temperatures (Stillwater Sciences, 2010). Before colonial occupation and settlement, the Wiyot tribe freely inhabited the lower part of the Baduwa't Watershed since it is included in their ancestral territory. The upper part of the watershed was inhabited by the Whilkut, Nogatl, and Lassik. (Stillwater Sciences, 2010). Today, the Wiyot-affiliated Blue Lake Rancheria is federally recognized and holds land in its name. In the 1980s, the Whilkut, Nogatl, and Lassik were extirpated during the Indian Wars (Stillwater Sciences, 2010). While each group inhabited this watershed, they had boundaries on land, resources, and management practices. They interacted with the river, its surrounding environment, and wildlife to create means of food, tools, medicines, spiritual community, and connectedness. According to the Mad River Watershed Assessment Final Report, during this time, "the region's vegetative communities are recognized as being among the most biologically rich, diverse and productive environments in the world" (Stillwater Sciences, 2010). The article mentioned the North Coast ranges during the pre-colonial times compared to our current time, "surpassed the population density of agricultural societies" (Stillwater Sciences, 2010).

In 1849 the discovery of gold brought white settlers to California. In their search for gold, the settlers eventually found their way to the Humboldt Bay region and Baduwa't Estuary. While they did not discover significant gold, they did find lands with vast forests and waters rich with many species of fish. The settlers exploited the land for resources through large-scale extraction of timber and fish. The land was cleared to make settlements and agricultural operations. Dams were constructed to provide water to both the settlements and farms which further altered the ecology of the river. Between 1860 and 1910 the population of the Wiyot decreased from approximately 1,500-2,000 to only 100 people due to a violent genocide at the hands of the settlers. Indigenous tribes in this area, such as the Wiyot, have had a strong understanding of the ecology of the land and fostered a relationship of stewardship in concert with it. Due to this, when their numbers were greatly reduced, the ecology of the land also suffered along with them. (Wiyot Tribe, n.d.). The West Coast alone has seen an 85% loss of tidal wetlands since the settlement of Europeans (Hodgson et al., 2020). This, coupled with the transition to agricultural lands in the 1980s, has severely altered the ecology of the Baduwa't Estuary and the lands surrounding it. Recently, the project's grading phase was completed at the end of 2022, and revegetation was done between November 2022 and March 2023. This project restored habitat for juvenile salmon by constructing an off-channel pond (Acosta et al., 2023).

The climate at the lower watershed of the Baduwa't Estuary is greatly affected by the Pacific Ocean's marine influences and it experiences 40-50 inches of rain every year. Precipitation primarily occurs between November and March. During the summer months, the high temperatures are commonly within the 50's and 60's in degrees Fahrenheit. During the winter, high temperatures tend to average in the low 40's to mid 50's. Winter low temperatures can reach as low as the mid 30's. The climate in the mid-upper watershed is characterized by hot and dry summers and cold and wet winters. Temperatures in summer average from 80-90 degrees Fahrenheit, reaching over 100 degrees at certain times. Average annual precipitation for the entire watershed is approximately 60 inches with up to 75 inches in high headwaters primarily falling between October and April (Stillwater Sciences, 2023).

In the 1980s, the area surrounding the project site was largely converted into agricultural lands and was impacted by various alterations such as flood control levees along the river and the construction of wastewater percolation ponds (California Trout, 2019). These changes caused a large decrease in biodiversity and habitat and came at a heavy ecological cost. This conversion to agricultural land also impacted the Baduwa't and Estuary by creating various agricultural levees that divert water. The focal point of this project is located in the Baduwa't Estuary and is within a tidal influence that is approximately 3 miles off the river. The Baduwa't is known for its salmonid populations and is home to the Coho, Steelhead, and Chinook Salmon, as well as resident trout species (California Trout, 2019). Salmon are largely regarded as a keystone species because they provide crucial nutrients from the sea that help provide sustenance for coastal forests and specialized vascular species such as redwood trees. Salmon are also a crucial food source for much of the wildlife in the surrounding areas. These salmon are culturally and spiritually significant to various tribes in the surrounding areas such as the Wiyot, Yurok, Tolowa, and Hupa tribes (Markos, 2004). The river drains 497 square miles over roughly 100 miles to the Pacific Ocean but is limited in estuaries and off-channel habitats to support the rearing of young fish. This makes the Baduwa't Estuary of particular interest from an ecological standpoint, as it is one of the only major rearing grounds for salmonids. The project has helped to increase biodiversity by creating a riparian off-channel habitat.

California Trout is the organization spearheading this project and they have extensively revegetated the area with caged plantings but saw only a 41.5% survival rate with 50% dead and 2.7% on the borderline (California Trout, 2024). The high amount of death was speculated to be from the summer heat and lack of access to groundwater since the elevation at this site varies. Survival rates were higher for caged plantings in shaded areas. The plants were caged to prevent herbivory and to increase survivorship rates. We wanted to understand the contributing factors to low survivorship rates. Since the site had varying elevations within its topography; we wanted to see if depth to groundwater would be an important factor for the placement of new plantings (Mcbain Associates, 2019). We hypothesize that elevation and root proximity to groundwater will significantly influence plant success rates in the Baduwa't Estuary. Our objectives are to explore the relationships between elevation, topography, and proximity to groundwater with native plant revegetation.

## Methods

### Site Description

The site is owned by McKinleyville Community Services District (MCSD) and is located in McKinleyville, California where School

Road and Ocean Drive intersect (Figure 1). We focused on an area with an off-channel pond in a riparian corridor. The most dominant vegetation species in this area are *Alnus rubra* (red alder), *Alopecurus* (foxtail), *Salix hookeriana (*Hooker's willow), and *Achillea millefolium* (yarrow). The dominant fish species are *Oncorhynchus kisutch* (coho salmon), *Petromyzontiformes* (lamprey), *Oncorhynchus mykiss* (Northern California steelhead), and *Oncorhynchus tshawytscha* (Chinook salmon) (California Trout, 2019). The Baduwa't Estuary was originally connected to a vast floodplain and was used by the Wiyot tribe (Burke and Gurin, 2023). Juvenile salmon were able to live harmoniously and thrived in the off-channel habitats of the estuary. However, European settlers significantly changed the floodplain to assist with flood control and agriculture, thus destroying many of the habitats actively used by salmon (Burke and Gurin, 2023).

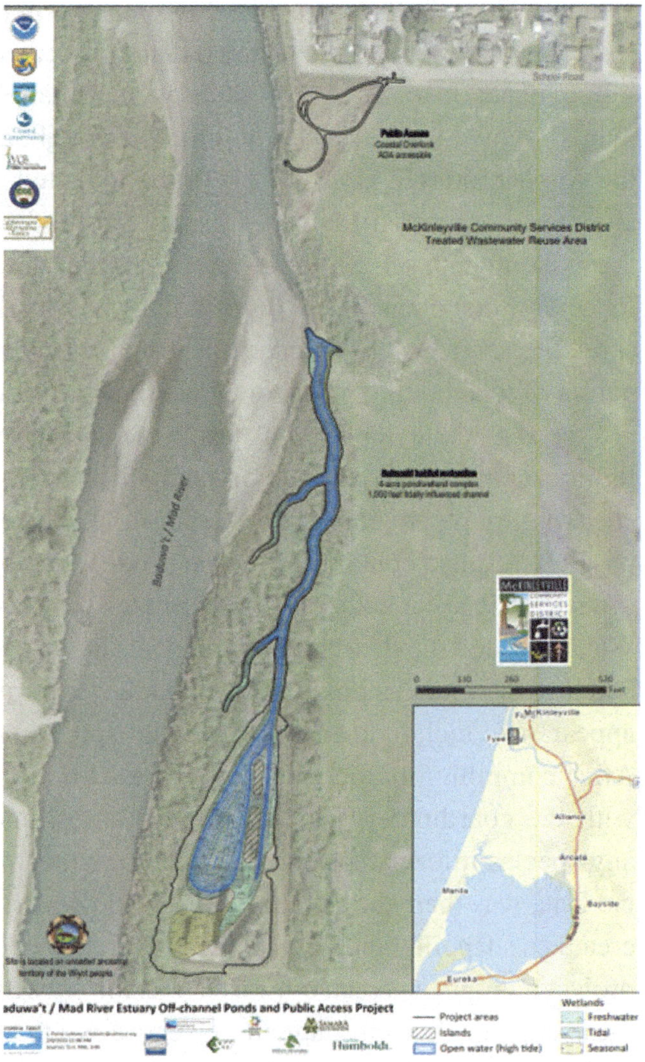


*Figure 1: Baduwa't/Mad River Estuary Off Channel Ponds and Public Access Project Map*

## Objectives

For this project, our objective was to create a visual representation of plant success over elevation. To do this we needed to assess and document the current life statuses of all plants at the project site and create an accurate raster file containing exact elevations throughout the site to use in ArcGIS Pro. We needed this information to be mapped out and placed within data tables that we could populate graphical representations of. The map would help us meet our objectives by allowing us to see exactly what elevations plants were dying at and the graphs would give us exact rates of success at these elevations. With this data, we would be able to assess whether there was any significant correlation between any particular elevation and plant survivorship rates.

## Field Methods

We utilized GIS data from the project's revegetation effort to test our hypothesis regarding plant success in proximity to groundwater (California

Trout, 2024). The geographic coordinate points were organized, and this data displays the perceived living, dead, or uncertain life statuses of all native plants involved in revegetation efforts throughout the project site (Figures 21-24). We conducted a site reassessment on March 17th, 2024, to update the existing data set. Multiple strategies were used to assess the plants' health including the "bark scratch test" which allowed for a visual determination of whether the plant contained moisture and produced a healthy color. This test is conducted by lightly scratching the outside of the main stem and checking for moisture and green color from chlorophyll. If the plant contains water and produces chlorophyll, it is alive and conducting photosynthesis. We also tested the flexibility of the plants by gently bending their main stems and checked their overall appearance, including for any potential sprouting. After compiling the new data, we created a table with the coordinates, updated plant statuses, and any species information that had been provided to us. Data was then entered into ArcGIS Pro and we created a shapefile with each plant as a marker. The project's engineer created a raster file for us with the exact elevations throughout the entire site to be used in GIS analysis (figure 2). The plants' statuses were then compared to their elevation to check for any correlations related to this limiting factor. If plants at higher elevations are more frequently not surviving, or plants at lower elevations are staying alive, this would prove that elevation had played a significant role in plant survivability. Due to irrigation challenges, it was of great importance for us to determine whether proximity to groundwater is the reason certain plantings have historically displayed low survivorship.

The first step in our process was to reassess the statuses of all the planted species at the site that were previously marked as "dead" or "maybe alive" for survivorship. We visited each of these plants by utilizing GPS coordinates that were provided to us through a GIS shapefile by Cal Trout that was based on their latest assessment. This file included exact latitude and longitude coordinates for each plant so that we could locate them. To see the full list of plant statuses and coordinates and their statuses, see figures 21-24 in the Appendix. After compiling a list of the statuses along with their GPS coordinates, we then utilized a topographic map of the site that the project's environmental engineer compiled. This map included all elevations in meters on a raster file (figure 2). By color-coding the plant coordinates by their life statuses and overlaying them on the topographic map, we were able to compare survivorship rates at various elevations throughout the site to determine if there was any clear correlation (Tables 2-19). Following this, we compiled the totals of status.

We took counts of all the "live" and "dead" plants along with their elevation and species to compile various data tables. We excluded all "maybe" statuses for scientific rigor, as these would not allow us to draw any clear conclusions. Two less specific groups were also created because there were members of the Salix and Roses genera that were not fully identified. They were grouped into their own tables, while all other unidentified species were placed in the "other" category. Finally, we took the totals of all life statuses by elevation and created a graph demonstrating the total number of live and dead plants in meters of elevation for comparison (figure 3).

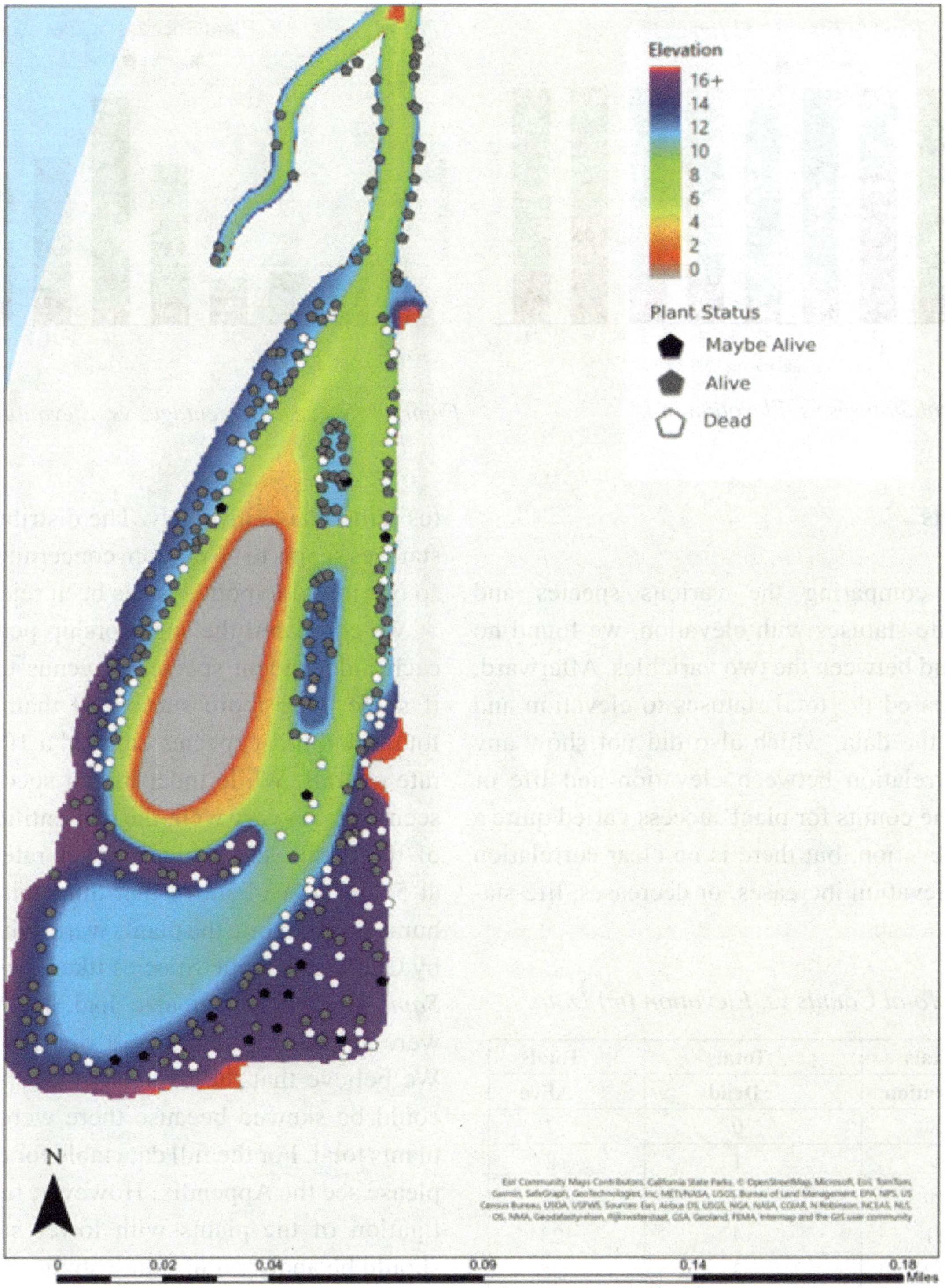


*Figure 2: Site Elevation (m) Map with GPS Coordinates by Status.*

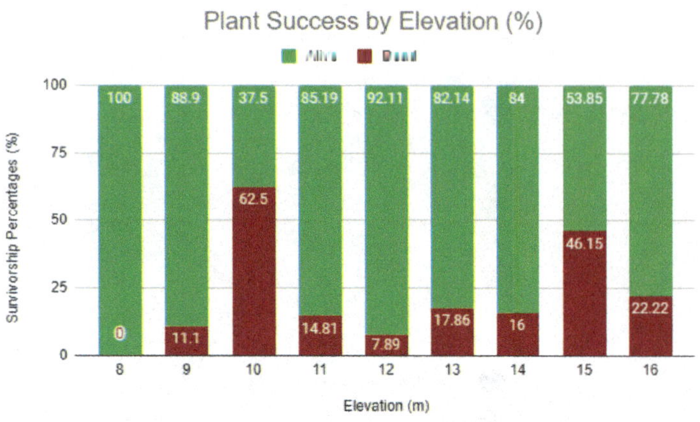


*Figure 3: Total Statuses vs. Elevation (m).*

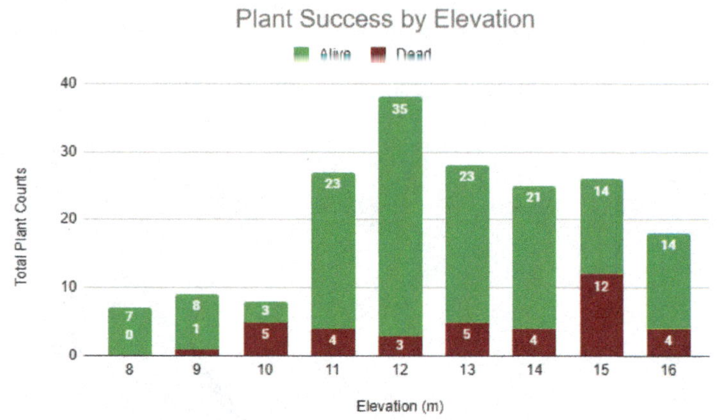


*Figure 4: Success Percentages vs. Elevation (m).*

**Results**

After comparing the various species and genera life statuses with elevation, we found no clear trend between the two variables. Afterward, we compared the total statuses to elevation and graphed the data, which also did not show any clear correlation between elevation and life or death. The counts for plant success varied quite a bit by elevation, but there is no clear correlation that as elevation increases, or decreases, life-status will behave inversely. The distribution of life statuses seems to be random concerning elevation, so our initial hypothesis has been rejected.

*Table 1: Total Counts vs. Elevation (m) Data.*

| Totals | Totals | Totals |
|---|---|---|
| **Elevation** | **Dead** | **Alive** |
| 8 | 0 | 7 |
| 9 | 1 | 8 |
| 10 | 5 | 3 |
| 11 | 4 | 23 |
| 12 | 3 | 35 |
| 13 | 5 | 23 |
| 14 | 4 | 21 |
| 15 | 12 | 14 |
| 16 | 4 | 14 |

We compared the survivorship percentages of each independent species or genus to determine if some were more successful than others. We found that most species enjoyed a 100% success rate overall. While independent species of Salix seemed to do very well, the unidentified members of the genus had a much lower rate of success at 55.56%. We believe that this could be due to human error while the plants were being identified by California Trout. Species like *Alnus rubra* and *Sambucus racemosa* also had death rates that were surprisingly high for a riparian ecosystem. We believe that the *Sambucus racemosa* results could be skewed because there were only three plants total. For the full data table for each species please see the Appendix. However, further investigation of the plants with lower success rates should be analyzed in future studies to determine attributing factors. Species-specific needs such as soil type, exposure to sunlight, root type, and other factors could be contributing to these variances.

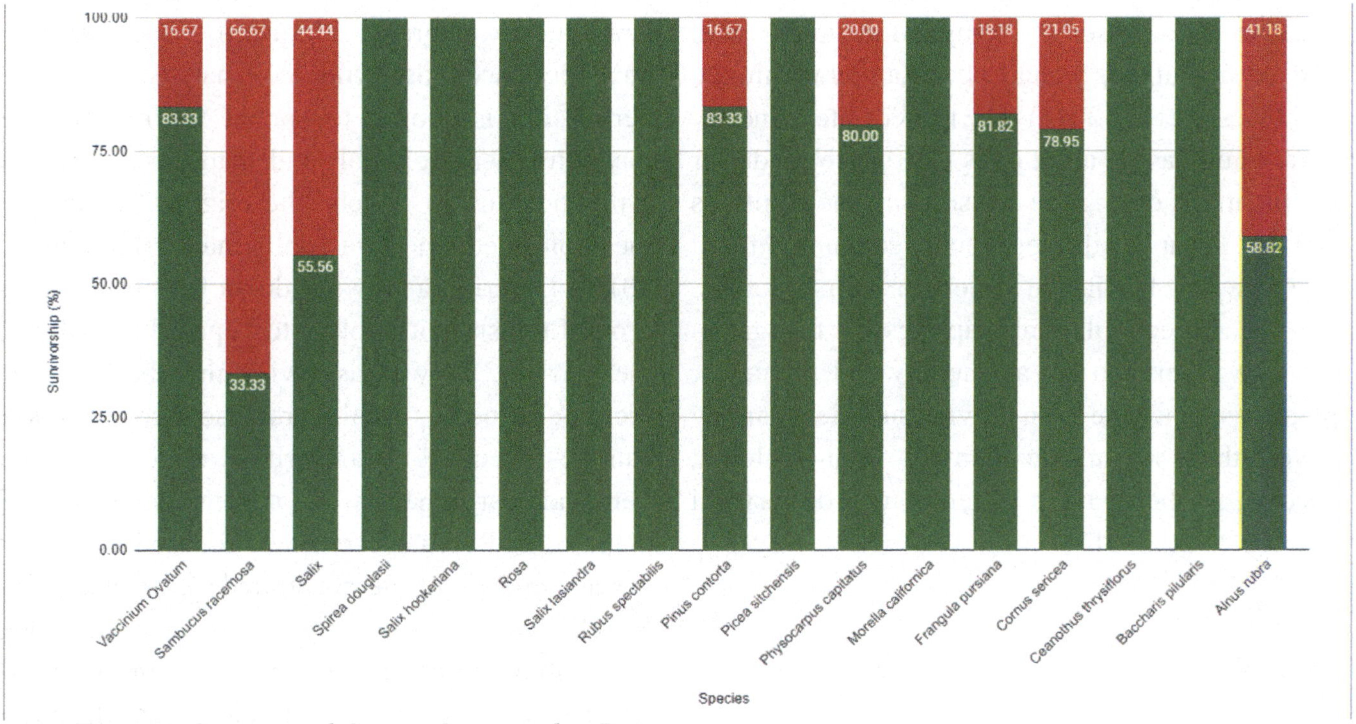


*Figure 5: Species and Genera Survivorship Percentages*

## Discussion

### Channel Access

The results show that higher elevations had a similar survivorship rate to lower elevations. We assumed that higher elevation would have lower survivorship since the roots of the plants would have a farther distance to travel to access the groundwater table. However, the plantings at higher elevations were just as successful. This may be because a significant number of the plantings at higher elevations were near the Baduwa't. Most plant's roots within a riparian corridor cannot reach further than a meter into the soil. Due to this, they rely on nearby channels, streams, or rivers during the summer months to supply them with the water necessary for survival (Dosskey et al., 2010). Since these plantings are within the riparian corridor, they can easily access the water from the Baduwa't even during the dry season. These species can also retain nutrients better than plants farther away from the riparian area since they can uptake groundwater and overland flow more readily (Entry and Emmingham, 1996). It would be best to plant riparian species (alders, willows, and cottonwoods) closer to the riparian corridor since they tend to have shorter root systems and are adapted to this kind of habitat.

### Canopy Cover

Some of the plantings did not survive, which could be due to a lack of canopy cover. Canopy

cover can increase plant growth and development because it decreases the temperature in the soil, and air, and allows water to be retained within the soil (Moore et al., 2011). The project site is mostly barren and has no adult trees to provide shade for the plantings during the dry season. The plantings can dry up and wilt from the direct sunlight and scarce water, leading to lower survivorship rates. We recommend planting saplings of fast-growing tree species to act as shelterwood for newer plantings. Once they can provide sufficient canopy cover, then we advise planting light-intolerant species, so they have a better chance of survival (Moore et al., 2011).

### Plant Care

Once a location is determined for revegetation, planting can be as simple as preparing a hole, placing the individual, and tucking back in the topsoil. Depending on the state of the plug, pot-grown seedling/ sapling, or cutting, extra consideration towards ensuring root success may prove to be very beneficial. If nursery or container conditions limit the amount of available space and soil, roots can easily become bound which reduces both their quality and chances of success in a new environment (Landis et al., 1990). Common practice includes massaging or breaking apart bound roots to increase the ability of the plant to settle into the soil for a long life ahead. The spectrum of this method ranges from gentle to potentially damaging.

Indigenous perspectives of plants as sacred relatives urge a slower process. Robin Wall Kimmerer writes about this considerate relationship in the Honorable Harvest chapter of her book Braiding Sweetgrass. Eloquently, she explains an experience post wild leek harvest in which permission is sought to spread the plants to a location where colonial land management led to a monoculture in which “the original habitat for these old medicines was obliterated” (Kimmerer, 2015). Describing slowing down before breaking ground to ask if it is okay to “open the earth for their arrival,” as well as envisioning the future of reciprocity begins to illustrate how revegetation can tie to culture and spirit (Kimmerer, 2015). Personal participation in several local native planting and invasive removal-focused volunteer stewardship events demonstrated a clear necessity for efficiency. Community involvement provides a valuable opportunity to restore connections to place, plants, and other people. However, education about methods must be thoroughly considered to ensure quality efforts alongside goals of quantity.

### Root Access

Plants need space to grow their roots, available soil to expand, and access to groundwater. While planting native plants in an area there are a few barriers that may come up. One of these barriers may be the contents of the soil. As we were planting native plants on our site, we encountered areas where there were concrete layers that hindered our ability to dig deep enough holes for plant stability and structure. Realizing this, plant growth in these areas may affect the way plants in the area grow. Our recommendations for combating inaccessible water resources would be through mulching and mycorrhizae association to create water retention.

Mulching is used for the protection of the soil from erosion and reduces water evaporation from the topsoil, which in turn can hold soil moisture for plants (Zhang et al., 2023). In terms of water-holding capacity, bran mulching is the most effective in absorbing water (Zhang et. al., 2023). Mycorrhizae is another possible way we can improve plant growth. Mycorrhizae association can also extend root surface area and allow plants to uptake water and nutrients more efficiently than before. Glomeromycota is compatible with 90% of plant species so it may be the best fit for an area with an abundance of different species (Holden et al., 1983).

Many variables can influence the survivorship of plants. Factors such as species adaptations, shifts in ecology, land use, and other environmental changes can all have a significant impact. Further investigation is needed into these various components to determine the exact causes of plant deaths on the site. As this project continues with plant revegetation, using these recommendations can create better survivorship rates in the future.

## Acknowledgments

We would like to extend our sincerest appreciation to the leaders who brought this project into our hands and minds: our mentors, Mary Burke of California Trout and Daniel Lipe (Bubba) of California State Polytechnic Humboldt. We are very grateful for the time and effort that they dedicated to ensuring we were supported every step of the way. Another special thanks go to Dr. Alison O'Dowd for helping us efficiently organize our data.

## Bibliography

Acosta, G., Bacon, P., & Corcoran, K. (2023). *Coastal Prairie Enhancement Recommendations for Lhiwetgut, the Baduwa't Estuary Overlook*. https://californiatroutinc.app.box.com/s/cvxtqwm4itc5ezx00eetrl8pe4sfrg0q/file/1386882966657

*Basic Information about Estuaries*. EPA. (n.d.). https://www.epa.gov/nep/basic-information-about-estuaries

Burke, M., & Gurin, K. (2023, February 16). *Restoring the baduwa't estuary*. California Trout. https://caltrout.org/current-article/restoring-baduwat

California Trout. (2024). *Vegetation Monitoring Report Rev. 1 Baduwa't Estuary Restoration Project*.https://californiatroutinc.app.box.com/s/cvxtqwm4itc5ezx00eetrl8pe4sfrg0q/file/1448409609387

Dosskey M., Vidon P., Gurwick N., Allan C., Duval T., Lowrance R. (2010). *The role of riparian vegetation in protecting and improving chemical water quality in streams*. Journal of the American Water Resources Association. https://doi.org/10.1111/j.1752-1688.2010.00419.x

Entry J and Emmingham W. (1996). *Nutrient content and extractability in riparian soils supporting forests and grasslands*. Elsevier. https://doi.org/10.1016/0929-1393(96)00111-4

Hodgson, E. E., Wilson, S. M., & Moore, J. W. (2020). Changing estuaries and impacts on juvenile salmon: A systematic review. *Global Change Biology*, *26*(4), 1986–2001. https://doi.org/10.1111/gcb.14997

Kimmerer, R. W. (2015). *Braiding Sweetgrass*. Milkweed Editions.

Landis, T. D., Tinus, R. W., McDonald, S. E., Barnett, J. P., Dumroese, R. K., Haase, D. L., Nisley, R. G., & Service, U. S. F. (1990). *The Container Tree Nursery Manual: Seedling processing, storage, and outplanting*. U.S. Department of Agriculture, Forest Service. https://books.google.com/books?id=AtyLKnWoSc8C

Markos, Sally. *Salmon and the Ecosystem*, City of Eugene Public Works, Mar. 2004, www.eugene-or.gov/DocumentCenter/View/3291/SalmonCurriculum5-8.

McBain Associates. (2019). *Redwood National and State Park Visitor Center and Restoration Project Basis of Revegetation Design Report Final Report*. Save the Redwoods League. https://californiatroutinc.app.box.com/s/cvxtqwm4itc5ezx00eetrl8pe4sfrg0q/file/1448403462752

Moore P., Holl K., Wood D. (2011). *Strategies for restoring native riparian understory plants along the sacramento river: timing, shade, non-native control, and planting method.* eScholarship. https://escholarship.org/uc/item/7555d3b4

Stillwater Sciences. (2010). *Mad River Watershed Assessment Final Report*. Redwood Community Action Agency and Natural Resources Management Corp. https://californiatroutinc.app.box.com/s/cvxtqwm4itc5ezx00eetrl8pe4sfrg0q/file/1335149689350?sb=/details

*History | Wiyot Tribe, CA.* (n.d.). Retrieved March 31, 2024, from https://www.wiyot.us/148/Cultural

Holden, J. M., Thomas, G. W., & Jackson, R. M. (1983). Effect of mycorrhizal inocula on the growth of Sitka spruce seedlings in different soils. *Plant and Soil*, *71*(1), 313–317. https://doi.org/10.1007/BF02182669

Zhang, P., Xiao, M., Zhang, Z., Dai, Y., Liu, G., & Senge, M. (2023). Hydrological characteristics of different organic materials mulches. *Scientific Reports*, *13*, 771. https:/ doi.org/10.1038/s41598-023-28124-y

## Appendix

*Table 2: Vaccinium ovatum Status vs. Elevation (m) Data.*

| Plant | Status | Elevation |
|---|---|---|
| 1 | Alive | 14 |
| 2 | Alive | 12 |
| 3 | Alive | 12 |
| 4 | Alive | 13 |
| 5 | Alive | 16 |
| 6 | Dead | 16 |

*Table 3: Sambucus racemosa Status vs. Elevation (m) Data.*

| Plant | Status | Elevation |
|---|---|---|
| 1 | Dead | 11 |
| 2 | Alive | 12 |
| 3 | Dead | 13 |

*Table 4: Unidentified Species Within the Genus Salix vs. Elevation (m) Data.*

| Plant | Status | Elevation |
|---|---|---|
| 1 | Alive | 8 |
| 2 | Alive | 9 |
| 3 | Alive | 11 |
| 4 | Dead | 9 |
| 5 | AJ1ve | 14 |
| 6 | Dead | 15 |

| Plant | Status | Elevation |
|---|---|---|
| 7 | Dead | 15 |
| 3 | Dead | 14 |
| 9 | Dead | 15 |
| 10 | Dead | 15 |
| 11 | Dead | 12 |
| 12 | Dead | 13 |
| 13 | Alive | 12 |
| 14 | Alive | 12 |
| 15 | Alive | 13 |
| 16 | Ahve | 15 |
| 17 | Dead | 15 |
| 18 | Alive | 14 |
| 19 | AJ1ve | 15 |
| 20 | Dead | 13 |
| 21 | Dead | 14 |
| 22 | Alive | 12 |
| 23 | Dead | 15 |
| 24 | Dead | 16 |
| 25 | Alive | 9 |
| 26 | AJ1ve | 9 |
| 27 | Alive | 9 |
| 28 | Ahve | 11 |
| 29 | Alive | 10 |
| 30 | Alive | 10 |
| 31 | Alive | 16 |
| 32 | Dead | 13 |
| 33 | AJ1ve | 9 |
| 34 | Dead | 10 |
| 35 | Dead | 10 |

*Table 5: Salix hookeriana vs. Elevation (m) Data.*

| Plant | Status | Elevation |
|---|---|---|
| 1 | Alive | 8 |
| 2 | Alive | 10 |
| 3 | Alive | 11 |

*Table 6: Spirea douglasii vs. Elevation (m) Data.*

| Plant | Status | Elevation |
|---|---|---|
| 1 | Alive | 8 |
| 2 | Alive | 9 |
| 3 | Alive | 11 |
| 4 | Alive | 12 |
| 5 | Alive | 8 |
| 6 | Alive | 9 |
| 7 | Alive | 13 |
| 8 | Alive | 11 |
| 9 | Alive | 11 |
| 10 | Alive | 11 |
| 11 | Alive | 14 |
| 12 | Alive | 12 |
| 13 | Alive | 13 |
| 14 | Alive | 11 |
| 15 | Alive | 8 |
| 16 | Alive | 11 |

*Table 7: Unidentified Species Within the Genus Rosa vs. Elevation (m) Data.*

| Plant | Status | Elevation |
|---|---|---|
| 1 | Alive | 13 |
| 2 | Alive | 13 |
| 3 | Alive | 11 |
| 4 | Alive | 11 |
| 5 | Alive | 11 |
| 6 | Alive | 11 |

*Table 8: Salix lasiandra vs. Elevation (m) Data.*

| Plant | Status | Elevation |
|---|---|---|
| 1 | Alive | 14 |
| 2 | Alive | 12 |
| 3 | Alive | 12 |
| 4 | Alive | 14 |
| 5 | Alive | 16 |
| 6 | Alive | 16 |
| 7 | Alive | 16 |

*Table 9: Rubus spectabilis vs. Elevation (m) Data*

| Plant | Status | Elevation |
|---|---|---|
| 1 | Alive | 12 |
| 2 | Alive | 12 |
| 3 | Alive | 12 |
| 4 | Alive | 14 |

*Table 10: Pinus contorta vs. Elevation (m) Data.*

| Plant | Status | Elevation |
|---|---|---|
| 1 | Alive | 12 |
| 2 | Alive | 12 |
| 3 | Alive | 12 |
| 4 | Alive | 12 |
| 5 | Alive | 12 |
| 6 | Alive | 12 |
| 7 | Alive | 12 |
| 8 | Alive | 13 |
| 9 | Dead | 16 |
| 10 | Alive | 16 |
| 11 | Alive | 13 |
| 12 | Dead | 16.5 |

*Table 11: Picea sitchensis vs. Elevation (m) Data.*

| Plant | Status | Elevation |
|---|---|---|
| 1 | Alive | 9 |
| 2 | Alive | 8 |
| 3 | Alive | 15 |
| 4 | Alive | 15 |
| 5 | Alive | 15 |
| 6 | Alive | 15 |
| 7 | Alive | 14 |
| 8 | Alive | 13 |
| 9 | Alive | 12 |
| 10 | Alive | 12 |
| 11 | Alive | 13 |
| 12 | Alive | 11 |
| 13 | Alive | 12 |
| 14 | Alive | 12 |

*Table 12: Physocarpus capitatus vs. Elevation (m) Data.*

| Plant | Status | Elevation |
|---|---|---|
| 1 | Dead | 15 |
| 2 | Alive | 11 |
| 3 | Alive | 11 |
| 4 | Alive | 13 |
| 5 | Alive | 12 |

*Table 13: Morella californica vs. Elevation (m) Data.*

| Plant | Status | Elevation |
|---|---|---|
| 1 | Alive | 8 |
| 2 | Alive | 12 |
| 3 | Alive | 12 |
| 4 | Alive | 12 |
| 5 | Alive | 13 |
| 6 | Alive | 13 |
| 7 | Alive | 16 |

*Table 14: Frangula pursiana vs. Elevation (m) Data.*

| Plant | Status | Elevation |
|---|---|---|
| 1 | Dead | 10 |
| 2 | Alive | 14 |
| 3 | Dead | 12 |
| 4 | Alive | 15 |
| 5 | Alive | 15 |
| 6 | Alive | 13 |
| 7 | Alive | 14 |
| 8 | Alive | 16 |
| 9 | Alive | 16 |
| 10 | Alive | 15 |
| 11 | Alive | 11 |

*Table 15: Cornus sericea vs. Elevation (m) Data.*

| Plant | Status | Elevation |
|---|---|---|
| 1 | Dead | 10 |
| 2 | Alive | 11 |
| 3 | Alive | 14 |
| 4 | Dead | 14 |
| 5 | Alive | 15 |
| 6 | Alive | 15 |
| 7 | Alive | 14 |
| 8 | Alive | 14 |
| 9 | Alive | 13 |
| 10 | Alive | 14 |
| 11 | Alive | 12 |
| 12 | Dead | 12 |
| 13 | Dead | 11 |
| 14 | Alive | 11 |
| 15 | Alive | 13 |
| 16 | Alive | 12 |
| 17 | Alive | 12 |
| 18 | Alive | 12 |
| 19 | Alive | 11 |

*Table 16: Ceanothus thrysiflorus vs. Elevation (m) Data.*

| Ceanothus thrysiflorus | Ceanothus thrysiflorus | Ceanothus thrysiflorus |
|---|---|---|
| **Plant** | **Status** | **Elevation** |
| 1 | Alive | 12 |
| 2 | Alive | 13 |
| 3 | Alive | 14 |
| 4 | Alive | 11 |

*Table 17: Baccharis pilularis vs. Elevation (m) Data.*

| Plant | Status | Elevation |
|---|---|---|
| 1 | Alive | 13 |
| 2 | Alive | 13 |
| 3 | Alive | 13 |
| 4 | Alive | 12 |
| 5 | Alive | 12 |
| 6 | Alive | 11 |
| 7 | Alive | 16 |
| 8 | Alive | 16 |
| 9 | Alive | 16 |
| 10 | Alive | 16 |
| 11 | Alive | 16 |

*Table 18: vs. Alnus rubra Elevation (m) Data.*

| Plant | Status | Elevation |
|---|---|---|
| 1 | Alive | 15 |
| 2 | Dead | 15 |
| 3 | Dead | 11 |
| 4 | Alive | 15 |
| 5 | Dead | 15 |
| 6 | Dead | 15 |
| 7 | Alive | 13 |
| 8 | Alive | 13 |
| 9 | Alive | 14 |
| 10 | Dead | 15 |
| 11 | Dead | 11 |
| 12 | Dead | 13 |
| 13 | Alive | 15 |
| 14 | Alive | 12 |
| 15 | Alive | 13 |
| 16 | Alive | 11 |
| 17 | Alive | 11 |

*Table 19: Other Species vs. Elevation (m) Data.*

| Plant | Status | Elevation |
|---|---|---|
| 1 | Dead | 10 |
| 2 | Alive | 14 |
| 3 | Alive | 12 |
| 4 | Alive | 14 |
| 5 | Alive | 14 |
| 6 | Alive | 14 |
| 7 | Dead | 15 |
| 8 | Dead | 14 |

*Table 20: Species Total Survivorship by Percentage*

| Species | Alive (%) | Dead (%) |
|---|---|---|
| Vaccinium Ovatum | 83.33 | 16.67 |
| Sambucus racemosa | 33.33 | 66.67 |
| Salix | 55.56 | 44.44 |
| Spirea douglasii | 100.00 | 0.00 |
| Salix hookeriana | 100.00 | 0.00 |
| Rosa | 100.00 | 0.00 |
| Salix lasiandra | 100.00 | 0.00 |
| Rubus spectabilis | 100.00 | 0.00 |
| Pinus contorta | 83.33 | 16.67 |
| Picea sitchensis | 100.00 | 0.00 |
| Physocarpus capitatus | 80.00 | 20.00 |
| Morella californica | 100.00 | 0.00 |
| Frangula pursiana | 81.82 | 18.18 |
| Cornus sericea | 78.95 | 21.05 |
| Ceanothus thrysiflorus | 100.00 | 0.00 |
| Baccharis pilularis | 100.00 | 0.00 |
| Alnus rubra | 58.82 | 41.18 |

*Figures 21-24: Coordinates and Status of All Plants on Site.*

| Type | Original Status | New Status 3/17/24 | Latitude | Longitude |
|---|---|---|---|---|
| | Dead | Dead | 40.92827764000 | -124.12822322000 |
| Frangula purshiana | Maybe | Dead | 40.92827253000 | -124.12756904000 |
| | Dead | Alive | 40.92826246000 | -124.12820815000 |
| | Dead | Alive | 40.92819303000 | -124.12823866000 |
| Salix sp | Maybe | Alive | 40.92816657000 | -124.12780162000 |
| | Dead | Dead | 40.92816294000 | -124.12825987000 |
| Salix sp | Maybe | Dead | 40.92814588000 | -124.12758542000 |
| Other | Dead | Alive | 40.92812364000 | -124.12779624000 |
| | Dead | Alive | 40.92810382000 | -124.12757736000 |
| Cornus sericea | Maybe | Dead | 40.92807640000 | -124.12757444000 |
| | Dead | Alive | 40.92807483000 | -124.12835964000 |
| Other | Maybe | Dead | 40.92803633000 | -124.12758881000 |
| | Dead | Dead | 40.92802513000 | -124.12833182000 |
| | Dead | Dead | 40.92800936000 | -124.12778625000 |
| Salix sp | Dead | Dead | 40.92800266000 | -124.12837389000 |
| Cornus sericea | Maybe | Alive | 40.92799954000 | -124.12758906000 |
| | Dead | Dead | 40.92798149000 | -124.12761584000 |
| | Dead | Dead | 40.92797882000 | -124.12837543000 |
| | Dead | Dead | 40.92797269000 | -124.12757229000 |
| Alnus rubra | Dead | Dead | 40.92796330000 | -124.12841363000 |
| | Dead | Dead | 40.92794729000 | -124.12757787000 |
| | Dead | Dead | 40.92794148000 | -124.12761027000 |
| Alnus rubra | Dead | Dead | 40.92793098000 | -124.12837069000 |
| | Dead | Dead | 40.92792147000 | -124.12758131000 |
| | Dead | Dead | 40.92791559000 | -124.12839767000 |
| | Dead | Dead | 40.92789719000 | -124.12763110000 |
| | Dead | Dead | 40.92789132000 | -124.12780800000 |
| | Dead | Dead | 40.92788690000 | -124.12779360000 |
| | Dead | Alive | 40.92787880000 | -124.12761091000 |
| | Dead | Dead | 40.92785246000 | -124.12844528000 |
| | Dead | Dead | 40.92783177000 | -124.12840585000 |
| | Dead | Dead | 40.92782967000 | -124.12846502000 |
| | Dead | Dead | 40.92782905000 | -124.12762480000 |
| | Dead | Dead | 40.92782564000 | -124.12775756000 |
| | Dead | Dead | 40.92781746000 | -124.12844136000 |

| Type | Original Status | New Status 3/17/24 | Latitude | Longitude |
|---|---|---|---|---|
| | Dead | Maybe | 40.92774977000 | -124.12780139000 |
| Vaccinium ovatum | Maybe | Alive | 40.92773533000 | -124.12761622000 |
| | Dead | Dead | 40.92773136000 | -124.12778163000 |
| Other | Dead | Dead | 40.92769312000 | -124.12779552000 |
| Picea sitchensis | Maybe | Alive | 40.92769083000 | -124.12760540000 |
| | Dead | Dead | 40.92768264000 | -124.12851474000 |
| | Dead | Dead | 40.92766317000 | -124.12757475000 |
| | Dead | Alive | 40.92765762000 | -124.12765457000 |
| Alnus rubra | Dead | Dead | 40.92765719000 | -124.12762366000 |
| | Dead | Dead | 40.92765286000 | -124.12788069000 |
| Cornus sericea | Maybe | Dead | 40.92763784000 | -124.12766187000 |
| Salix sp | Dead | Dead | 40.92763254000 | -124.12797061000 |
| Salix sp | Dead | Dead | 40.92762932000 | -124.12803758000 |
| | Dead | | 40.92762710000 | -124.12757658000 |
| Physocarpus capitatus | Dead | Dead | 40.92761997000 | -124.12842605000 |
| Salix sp | Maybe | Alive | 40.92760507000 | -124.12837039000 |
| Salix sp | Dead | Alive | 40.92759986000 | -124.12829318000 |
| Other | Dead | Alive | 40.92759853000 | -124.12808391000 |
| | Dead | Alive | 40.92759242000 | -124.12771147000 |
| Other | Dead | Alive | 40.92758788000 | -124.12802348000 |
| Pinus contorta | Dead | Dead | 40.92758659000 | -124.12844509000 |
| Salix sp | Dead | Dead | 40.92757525000 | -124.12833059000 |
| Frangula purshiana | Dead | Dead | 40.92756989000 | -124.12774005000 |
| Other | Dead | Dead | 40.92756824000 | -124.12802054000 |
| Cornus sericea | Maybe | Alive | 40.92756710000 | -124.12758986000 |
| Other | Dead | Alive | 40.92755984000 | -124.12805648000 |
| Other | Dead | Alive | 40.92755912000 | -124.12837499000 |
| Alnus rubra | Dead | Dead | 40.92755328000 | -124.12814211000 |
| | Dead | Dead | 40.92754469000 | -124.12768615000 |
| Alnus rubra | Dead | Dead | 40.92754267000 | -124.12779191000 |
| Salix sp | Dead | Dead | 40.92754154000 | -124.12774051000 |
| | Dead | Dead | 40.92754078000 | -124.12756332000 |
| Salix sp | Dead | Dead | 40.92753601000 | -124.12812019000 |
| Other | Dead | Dead | 40.92753326000 | -124.12855182000 |
| Salix sp | Dead | Dead | 40.92752279000 | -124.12821573000 |
| | Dead | Dead | 40.92752116000 | -124.12779528000 |
| Salix sp | Dead | Dead | 40.92751482000 | -124.12765470000 |

| Type | Original Status | New Status 3/17/24 | Latutude | Longitude |
|---|---|---|---|---|
| Frangula purshiana | Dead | Dead | 40.92751099000 | -124.12776863000 |
| Salix sp | Dead | Dead | 40.92750987000 | -124.12772232000 |
| Salix sp | Dead | Dead | 40.92748714000 | -124.12779196000 |
| Salix sp | Maybe | Dead | 40.92748685000 | -124.12755418000 |
| | Dead | Dead | 40.92748020000 | -124.12761384000 |
| Salix sp | Dead | Dead | 40.92745755000 | -124.12786668000 |
| | Dead | Dead | 40.92743788000 | -124.12764208000 |
| | Dead | Dead | 40.92743236000 | -124.12758233000 |
| | Dead | Dead | 40.92742748000 | -124.12786855000 |
| Other | Dead | Dead | 40.92741796000 | -124.12858942000 |
| Frangula purshiana | Maybe | Alive | 40.92741241000 | -124.12773857000 |
| | Dead | Dead | 40.92739816000 | -124.12785564000 |
| Alnus rubra | Dead | Alive | 40.92739774000 | -124.12770150000 |
| Salix sp | Dead | Dead | 40.92738342000 | -124.12789866000 |
| | Dead | Dead | 40.92738188000 | -124.12780109000 |
| | Dead | Dead | 40.92736071000 | -124.12792218000 |
| | Dead | Dead | 40.92735822000 | -124.12770209000 |
| Cornus sericea | Maybe | Alive | 40.92735814000 | -124.12758534000 |
| | Dead | Maybe | 40.92733554000 | -124.12781800000 |
| | Dead | Alive | 40.92732725000 | -124.12794795000 |
| Salix sp | Dead | Alive | 40.92732562000 | -124.12858413000 |
| | Dead | Dead | 40.92731946000 | -124.12790581000 |
| | Dead | Alive | 40.92731007000 | -124.12798146000 |
| | Dead | Dead | 40.92730976000 | -124.12776538000 |
| | Dead | Maybe | 40.92729498000 | -124.12757352000 |
| Alnus rubra | Dead | Dead | 40.92729272000 | -124.12771819000 |
| | Dead | Alive | 40.92729013000 | -124.12799388000 |
| Other | Dead | Alive | 40.92728644000 | -124.12864010000 |
| | Dead | Maybe | 40.92726623000 | -124.12792997000 |
| Other | Dead | Maybe | 40.92726327000 | -124.12785649000 |
| Other | Dead | Alive | 40.92724686000 | -124.12807980000 |
| | Dead | Alive | 40.92724674000 | -124.12796639000 |
| Cornus sericea | Maybe | Alive | 40.92724293000 | -124.12803208000 |
| Cornus sericea | Maybe | Dead | 40.92722415000 | -124.12860045000 |
| | Dead | Alive | 40.92721737000 | -124.12774578000 |
| Fran ula purshiana | Dead | Maybe | 40.92720960000 | -124.12809232000 |

| Type | Original Status | New Status 3/17/2024 | Latitude | Longitude |
|---|---|---|---|---|
| | Dead | Alive | 40.92720343000 | -124.12763800000 |
| Other | Dead | Alive | 40.92720151000 | -124.12780454000 |
| Alnus rubra | Dead | Alive | 40.92719456000 | -124.12815391000 |
| Other | Dead | Alive | 40.92718608000 | -124.12795730000 |
| Other | Dead | Maybe | 40.92718152000 | -124.12789095000 |
| Other | Dead | Alive | 40.92718145000 | -124.12807072000 |
| | Dead | Dead | 40.92717351000 | -124.12764262000 |
| | Dead | Dead | 40.92717334000 | -124.12769867000 |
| Vaccinium Ovatum | Maybe | Dead | 40.92717176000 | -124.12779299000 |
| Alnus rubra | Dead | Alive | 40.92715710000 | -124.12820698000 |
| | Dead | Alive | 40.92715451000 | -124.12769712000 |
| Other | Dead | Maybe | 40.92715288000 | -124.12797815000 |
| | Dead | Dead | 40.92714573000 | -124.12763253000 |
| Other | Dead | Alive | 40.92714523000 | -124.12785693000 |
| | Dead | Dead | 40.92714056000 | -124.12760381000 |
| Other | Dead | Dead | 40.92713235000 | -124.12819823000 |
| | Dead | Dead | 40.92713171000 | -124.12767882000 |
| Sambucus racemosa | Maybe | Dead | 40.92713159000 | -124.12862443000 |
| Alnus rubra | Dead | Dead | 40.92712432000 | -124.12801806000 |
| Pinus contorta | Maybe | Dead | 40.92711879000 | -124.12809241000 |
| | Dead | Dead | 40.92711716000 | -124.12769035000 |
| Alnus rubra | Dead | Maybe | 40.92711438000 | -124.12830006000 |
| Pinus contorta | Maybe | Maybe | 40.92710394000 | -124.12838787000 |
| Other | Dead | Dead | 40.92710032000 | -124.12847027000 |
| Other | Dead | Dead | 40.92709369000 | -124.12819856000 |
| Other | Dead | Dead | 40.92708947000 | -124.12814937000 |
| Other | Dead | Dead | 40.92706873000 | -124.12827249000 |

| Type | Original Status | New Status 3/17/24 | Latitude | Longitude |
|---|---|---|---|---|
| Salix sp | Dead | N/A | 40.93152283000 | -124.12713490000 |
| Salix sp | Dead | N/A | 40.93149989000 | -124.12713261000 |
| Frangula purshiana | Maybe | N/A | 40.93125887000 | -124.12699145000 |
| | Dead | N/A | 40.93074317000 | -124.12698609000 |
| Salix spitchensis | Maybe | N/A | 40.93068822000 | -124.12694826000 |
| Salix sp | Maybe | N/A | 40.93068387000 | -124.12694469000 |
| Salix hookeriana | Maybe | N/A | 40.93023333000 | -124.12713870000 |
| | Dead | N/A | 40.93022387000 | -124.12711304000 |
| Salix hookeriana | Maybe | N/A | 40.93019223000 | -124.12706781000 |
| | Dead | Dead | 40.92893044000 | -124.12772074000 |
| | Dead | Dead | 40.92891683000 | -124.12772833000 |
| | Dead | Alive | 40.92888194000 | -124.12756127000 |
| | Dead | Dead | 40.92886341000 | -124.12781310000 |
| | Dead | Dead | 40.92884851000 | -124.12778865000 |
| | Dead | Dead | 40.92883248000 | -124.12780046000 |
| | Dead | Dead | 40.92882622000 | -124.12754020000 |
| | Dead | Dead | 40.92880289000 | -124.12782232000 |
| | Dead | Alive | 40.92877883000 | -124.12784360000 |
| | Dead | Alive | 40.92873586000 | -124.12787746000 |
| | Dead | Alive | 40.92872328000 | -124.12753943000 |
| Sambucus racemosa | Maybe | Dead | 40.92871692000 | -124.12791569000 |
| Salix hookeriana | Dead | Alive | 40.92868666000 | -124.12754413000 |
| | Dead | Dead | 40.92866863000 | -124.12792433000 |
| | Dead | Dead | 40.92866468000 | -124.12754219000 |
| Rubus spectabilis | Maybe | Alive | 40.92864623000 | -124.12770150000 |
| | Dead | Dead | 40.92863587000 | -124.12754636000 |
| | Dead | Alive | 40.92862811000 | -124.12796220000 |
| | Dead | Dead | 40.92861068000 | -124.12755525000 |
| | Dead | Dead | 40.92860727000 | -124.12799992000 |
| | Dead | Dead | 40.92859286000 | -124.12797477000 |
| | Dead | Alive | 40.92859105000 | -124.12775703000 |
| Cornus sericea | Dead | Dead | 40.92856578000 | -124.12775387000 |
| | Dead | Alive | 40.92855956000 | -124.12755623000 |
| Salix sp | Maybe | Dead | 40.92854606000 | -124.12800914000 |
| | Dead | Dead | 40.92852206000 | -124.12776321000 |
| | Dead | Dead | 40.92851357000 | -124.12756201000 |

| Type | Original Status | New Status 3/17/24 | Latitude | Longitude |
|---|---|---|---|---|
| | Dead | Maybe | 40.92851327000 | -124.12768105000 |
| Salix sp | Maybe | Alive | 40.92850044000 | -124.12775748000 |
| Salix sp | Maybe | Dead | 40.92849203000 | -124.12804779000 |
| | Dead | Alive | 40.92848826000 | -124.12755525000 |
| Spiraea douglasii | Maybe | Alive | 40.92848065000 | -124.12807841000 |
| | Dead | Maybe | 40.92844879000 | -124.12806351000 |
| | Dead | Alive | 40.92844441000 | -124.12771323000 |
| | Dead | Dead | 40.92842329000 | -124.12808217000 |
| | Dead | Alive | 40.92842195000 | -124.12811585000 |
| | Dead | Dead | 40.92841747000 | -124.12768765000 |
| | Dead | Dead | 40.92840604000 | -124.12756553000 |
| | Dead | Maybe | 40.92837786000 | -124.12756315000 |
| | Dead | Alive | 40.92835258000 | -124.12814181000 |
| | Dead | Dead | 40.92835148000 | -124.12755610000 |
| Alnus rubra | Dead | Alive | 40.92833951000 | -124.12820608000 |
| | Dead | Dead | 40.92832968000 | -124.12756992000 |
| | Dead | Dead | 40.92830325000 | -124.12817926000 |
| | Dead | Alive | 40.92830132000 | -124.12758092000 |

IdeaFest
Journal

# Rending the Social Fabric: Vigilante Strikebreaking in 1930s California

**Abigail** Claybrook

On August 21 of 1935, at around one in the morning, several hundred vigilantes drove into a Jewish-American chicken rancher community in Petaluma in pursuit of labor organizers Jack Green and Sol Nitzberg. Green took shelter in Nitzberg's house, and the latter brought out a shotgun. The household remained in a standoff with the crowd for several hours, until the vigilantes launched tear gas through a window and hauled Green and Nitzberg into a car. The crowd brought Green, Nitzberg, and several other local labor organizers (this group had been organizing apple orchard workers in nearby Sebastopol in the weeks leading up to the event) to a warehouse in Santa Rosa. From there, vigilantes interrogated and beat the organizers, then singled out Green and Nitzberg. The men were stripped, tarred, feathered, and hauled down Fourth Street in a grotesque parade.[1]

An observer and member of the International Labor Defense (ILD) later asserted in an interview that "this thing was well planned, with the full approval of Sonoma County leaders. These vigilante actions were happening all over California as capitalists tried to prevent workers-from organizing trade unions and winning a living wage."[2] Sol Nitzberg's wife, Millie Nitzberg,-claimed that "a lot of the vigilantes were people of high standing from the American Legion, the Chamber of Commerce, the Republican party."[3] Perhaps most unsettling of all, Sol Nitzberg claimed to see Silver Shirts among the vigilante crowd.[4] The Silver Legion of America was a pro-fascist organization active through the 1930s in the United States.[5] The episode revealed a troubling collaboration between extremist elements and local business, all in the interest of terrorizing labor organizers.

1 Kenneth L. Kann, *Comrades and Chicken Ranchers: The Story of a California Jewish Community* (Ithaca: Cornell University Press, 1993), 105-107.
2 Kann, *Comrades and Chicken Ranchers*, 110.
3 Kann, *Comrades and Chicken Ranchers*, 111.
4 Kann, *Comrades and Chicken Ranchers*, 104.
5 Nelson A. Pichardo and Brian W. Kulik, *American Fascism and the New Deal: The Associated Farmers of California and the Pro-Industrial Movement* (Lanham: Lexington Books, 2013), 143.

This was no anomaly. U.S. labor endured all manner of aggressive strikebreaking for most of its existence. Labor historian Philip Taft once claimed that "the United States has had the bloodiest and most violent labor history of any industrial nation in the world."[6] During the 1930s, however, developments in labor and strikebreaking received particularly close documentation. According to labor historian Philip Taft, the New Deal era (1933-1938) was "the first time in peacetime history [that] union organizations had the attention and approval of the Federal Government."[7] Specifically, the Roosevelt administration oversaw the passage of the 1935 Wagner Act, as well as the LaFollette Committee (a civil liberties inquiry into violations of workers' free speech and assembly).8 So although the worker and employer remained in a bitter clash with one another, there existed a pro-labor coalition with federal backing throughout the decade.9 This coalition had the means and motivation to record and denounce antilabor violence.

Moreover, the methods of strikebreaking had changed by the end of the First World War. In the late nineteenth century, employers relied on the private detective to undermine labor. The interwar period (1918-1939), however, saw a transition from the hired strikebreaker to a "multifarious entity comprised of patriotic and veterans' organizations, law enforcement, military intelligence, employers' associations, and labor spies."[10] The Pinkerton faded away, replaced by localized vigilante organizations ultimately connected to a few overlapping corporate interests. With these factors in mind, the Sonoma mob was not an organic response to labor so much as it was the outcome of a network which had been years in the making.

Vigilantism was the speartip of California's far-flung anti-labor network. Central and Southern California constituted the epicenter of this network; key organizations fanned out from these regions and initially concentrated their efforts against union drives and strikes within the agricultural sector. With time, however, the network and its outgrowth of vigilante associations transcended regional and industrial divisions. Although it expanded well beyond Central and Southern California, the LaFollette Senate Report and subsequent accounts tend to devote their focus to that Central-Southern epicenter. Supplementing the traditional understanding of California's antilabor network with an examination of instances of antilabor violence in the north conveys its scope and pervasiveness.

Strikebreaking tactics and vigilante violence pioneered in the south carried over to Northern California and endangered not only labor but the very social fabric of the region, given the looming threat of reactionary violence. A full explanation calls for an examination of California's industrial and agricultural landscape, the circumstances which necessitated the LaFollette Committee investigation, the committee's findings, the network of antilabor organizations (with an emphasis on the Associated Farmers of California) present in California, vigilante bodies affiliated with

---

6 Philip Ross and Philip Taft, "American Labor Violence: Its Causes, Character, and Outcome," *The History of Violence in America: A Report to the National Commission on the Causes and Prevention of Violence*, ed. Hugh Davis Graham and Ted Robert Gurr (New York: F.A. Praeger, 1969), 280.

7 Ross and Taft, "American Labor Violence: Its Causes, Character, and Outcome," 347.

8 Jerold S. Auerbach, "The La Follette Committee: Labor and Civil Liberties in the New Deal," *The Journal of American History* 51, no. 3 (1964): 440, https://doi.org/ 10.2307/1894895.

9 Dolores E. Janiewski, "Through a Glass, Darkly: The NLRB, Employer Counteroffensives, Investigative Committees, and the CIO," In *Against Labor: How U.S. Employers Organized to Defeat Union Activism* (Champaign, IL: University of Illinois Press, 2017), 129, http://www.jstor.org/stable/10.5406/j.ctt1p6qq3f.9.

10 Simon J. Judkins, "Under Prying Eyes: Repression, Surveillance, and Exposure in California, 1918-1939," Master's thesis (Victoria University of Wellington, 2014), 2.

the network, and a survey of vigilante activity in Northern California from the Bay Area to the Oregon border.

## Vigilantism and the Conditions of Californian Labor

This discussion necessitates an overview of California's economy and history of vigilantism going into the 1930s. By the late nineteenth century, Californian agriculture had assumed a highly monopolized, industrial character. According to historian Donald Pisani, "an 1872 California legislative report showed that each of 122 individuals owned more than 20,000 acres in California."[11] While most Californian farmers were small farmers, those large landholders possessed most of the acreage and consequently dominated the agricultural sector. Corporations had a considerable hand in Californian agriculture. By 1939, Bank of America owned half of all farms in Central and Northern California alone.[12] The crops cultivated on this land were not meant for subsistence. They were perishable commodities. As such, large-scale growers in California could devote huge swathes of land to just one crop. They had the means to pursue a highly intensive mode of agricultural production wherein wage laborers worked compartmentalized tasks.[13] New Deal administration researcher and consultant Paul Taylor writes that this method of production "[gave] the state an agricultural proletariat, practically landless and propertyless, and exceedingly mobile."[14] Additionally, the sheer scale and centrality of California's agricultural sector meant that a whole host of other sectors were drawn into its orbit. Researchers Nelson Pichardo and Brian Kulik count "finance, power, packing, and transportation companies" among them.[15] All of these industries had a stake in agriculture's continued success and, by extension, a system of production which necessitated profound exploitation of its workers.

Turning back to the workers, this "agricultural proletariat" had historically been victimized by the state's longstanding tradition of vigilantism. Californian vigilantism dates back to the 1850s at least, and has assumed a racist and xenophobic character from its inception.[16] Such sentiments frequently commingled with economic anxieties, as well as a simultaneous desire to tamp down organized labor. Anti-Chinese rhetoric, for instance, conceived of Chinese workers as an impediment to white workers. During economic downturns, the rate of lynchings and expulsions rose dramatically. California saw renewed anti-Chinese violence during the 1870s and 1890s, for instance.[17] California's nonwhite minorities (including Japanese, Chinese, Mexican, Filipino, Indigenous, and Indian people) were vulnerable to vigilantism.[18] Employers only valued them insofar as they could be exploited. Researchers Mike Davis and Justin Akers Chacón note that "in face of the caste-like segmentation of the American working class, [it] is [not] very profitable to attempt to rigorously distinguish ethno-religious and racial violence from class violence."[19] This is all to say that vigilan-

11 Donald J. Pisani, "Land Monopoly in Nineteenth-Century California," *Agricultural History* 65, no. 4 (1991): 16, http://www.jstor.org/stable/3743944.

12 Carey McWilliams, *Factories in the Field* (Boston: Little, Brown and Company, 1939), 233.

13 Pichardo and Kulik, *American Fascism and the New Deal*, 32.

14 Paul Taylor, *On the Ground in the Thirties* (Layton: Peregrine Smith Books, 1983), 17.

15 Pichardo and Kulik, *American Fascism and the New Deal*, 31.

16 Justin Akers Chacón and Mike Davis, *No One is Illegal: Fighting Violence and State-Repression on the U.S.-Mexico Border* (Chicago, Illinois: Haymarket Books, 2006), 13.

17 Chacón and Davis, *No One is Illegal*, 27-29.

18 Chacón and Davis, *No One is Illegal*, 13.

tism was already well-established in the state of California. It was sporadic and typically localized. The antilabor network of the 1930s, however, provided infrastructure that would allow this violence to proliferate.

## The New Deal and the Network

The LaFollette Reports (compiled from hearings conducted between 1936 and 1940) shed light on strikebreaking during the New Deal era. In 1935, Wisconsin Senator Robert LaFollette Jr. called for a civil liberties inquiry into infringements on labor's right to free speech and assembly.[20] This call came at a time when labor's chief organ (the National Labor Relations Board, or NLRB) had been rendered inert by a glut of injunction proceedings, and could not carry out its stated functions.

The Roosevelt administration oversaw the passage of the 1935 National Labor RelationsAct (also known as the Wagner Act), which granted workers the right to organize unions, join unions, and bargain with their employer as a collective at the federal level. It also established the National Labor Relations Board (NLRB), meant to enforce this act. Employer associations across the country, anticipating strikes and union drives, mobilized against the Wagner Act. Many of these employers claimed that the act was unconstitutional, and filed suits against it. By the mid-1930s, the NLRB was paralyzed by these legal proceedings. The LaFollette Committee became labor's chief weapon against hostile employers. NLRB staff jumped ship, becoming field investigators and consultants for the committee.[21] Subsequent testimony compiled by committee investigators revealed that, at the national level, employers resorted to infiltration, crowd-control modes of violence (deploying tear gas, for instance), private policing, and hired strikebreakers to undermine labor.[22]

While assessing strikebreaking tactics, the LaFollette Committee paid special attention to the 1936 Mohawk Valley Formula. This formula is valuable in parsing antilabor rhetoric and strategy. Articulated by the president James Rand Jr. of the Remington Rand manufacturing company in the National Association of Manufacturers' (NAM) *Labor Relations Bulletin*, the Mohawk Valley Formula was a multi-step guide to strikebreaking. In the event of a strike, an employer would disseminate propaganda creating the impression that the union constituted an outside agitator. Once a strike began, the employer could escalate this claim by arguing that the outsider union was a threat to the local community. The employer would back public meetings known as "citizens committees," which would parrot this rhetoric and call for a forceful intervention. That intervention took the form of heavy policing and vigilante militias. All the while, employers maintained a stream of propaganda casting the union as an ineffectual minority and priming employees for an inevitable return to the workplace. If the strikers did not cave, law enforcement and militias menaced them. Rand explicitly encouraged forming a "warlike camp" against labor and its supporters.[23]

Throughout the process, the company set the terms of public discussion. It fostered a body of antilabor rhetoric, constructed a controlled

19 Chacón and Davis, *No One is Illegal*, 13.
20 Auerbach, "The La Follette Committee," 440.21 Auerbach, "The La Follette Committee," 439.
22 Auerbach, "The La Follette Committee," 439.
23 Ahmed A. White, "Industrial Terrorism and the Unmaking of the New Deal," *Nevada Law Journal*, 11, no. 561 (2011): 574.

opposition, and threatened violence against its opponent. Many prominent episodes of strikebreaking in California's history mirror or at least include many of the elements laid out in Rand's article. The Mohawk Valley Formula clarifies the strangely uniform nature of antilabor crackdowns and press coverage through the 1930s in the United States.

The LaFollette Committee struggled to secure funding for its entire duration, and lagged in carrying out an investigation of California as a result.[24] When it finally did, the 1940 Senate Report on the state "found belligerent [anti-labor] associations ... more active and of greater importance than elsewhere" in the country.[25] The concluding section argued that "this report on its California inquiry reflects basic problems of a nation-wide incidence, which are now and in the future will be the vital concern both of the Congress and the people," and warned that failure to address these developments in Californian industry would produce "either internal decay or the tyranny of small groups vested with great power."[26] The report drew attention to major strikebreaking entities such as the San Francisco Industrial Association and the Associated Farmers, both of which were dominated by corporate interests.

An examination of Associated Farmers (AF) is particularly instructive, as its connections and tactics were reflective of the broader California strikebreaking network. In 1933, in response to waves of agricultural strikes in the Imperial Valley, the California State Chamber of Commerce and California Farm Bureau Federation (both of which served as lobbying groups and employer associations) launched a joint committee assessing the condition of labor in agriculture.[27] This assessment culminated in the foundation of AF in 1934.[28] Members of the founding chapter in Imperial Valley vowed to: "help one another in case of emergency. They agreed to cooperate to harvest crops in case of strikes and to offer their services to the local sheriff immediately as special deputies in the event of disorders arising out of picketing and sabotage."[29] Although the organization derived from a cluster of corporate interests, it sought to cultivate a salt-of-the-earth appearance. During a meeting in May of 1934, AF representatives opted to make farmers into the face of the organization.[30] According to the 1940 Senate Report, AF's motivations boiled down to "opposition to unions of field-workers; opposition to unions of packinghouse and processing-house workers; and opposition to unions in industry."[31] The organization initially received its funding from Pacific Gas & Electric (PG&E) and California Packing Corporation (CalPak), both of which were vast corporate entities, not some fictitious coalition of yeoman farmers.[32] Its agricultural constituents were about as moneyed as PG&E and CalPak. Agricultural corporations and growers' associations sent donations in the tens of thousands of dollars to AF over the course of its existence.

In 1935, when AF was low on funds, the Industrial Association of San Francisco kept it afloat with donations.[33] AF's prolonged existence

24 Auerbach, "The La Follette Committee," 452.
25 *Violations of Free Speech and Rights of Labor. Report of the Committee on Education and Labor, Pursuant to S. Res. 266 Seventy-Fourth Congress, a Resolution to Investigate Violations of the Right of Free Speech and Assembly and Interference With the Right of Labor* (Washington D.C.: United States Government Printing Office, 1942), 7.
26 *Violations of Free Speech and Rights of Labor Report*, 62.
27 Taylor, *On the Ground in the Thirties*, 134.
28 Pichardo and Kulik, *American Fascism and the New Deal*, 67.
29 McWilliams, *Factories in the Field*, 231.
30 McWilliams, *Factories in the Field*, 232.
31 *Violations of Free Speech and Rights of Labor Report*, 46.
32 McWilliams, *Factories in the Field*, 232.

benefited both organizations; Congress passed the Wagner Act in July of that year, and employer's associations anticipated a renewed union drives across industries.[34] A robust anti-labor infrastructure was therefore imperative. Within the first year of its founding, representatives from AF had established chapters in twenty-six counties.[35] By 1939, it had a membership of forty thousand, with chapters in almost every county in the state of California.[36]

AF sought to alter local legislation and establish vigilante organizations to keep labor in check if the legal route failed. Anti-picketing ordinances became the first line of defense. AF chapters would push for legislation which directly or indirectly outlawed picketing. An ordinance could ban parades or some other form of assembly which happened to encompass the picket line.[37] This gave local law enforcement and their vigilante auxiliary forces carte blanche to brutalize and arrest picketers. Violating an ordinance could have escalating consequences for *labor organizers. The 1940 Senate Report on California explained that in an area operating* under an anti-picketing ordinance, "while picketing is only a misdemeanor, conspiracy is a felony and the charge of conspiracy has resulted in threatening the imposition of serious penalties on strike or union leaders."[38] Thus, labor organizers could be intimidated or imprisoned at length.

As previously mentioned, chapters of AF pledged themselves to local law enforcement. In the event of a strike, a sheriff could deputize hundreds of AF members in the name of public order. These local branches became paramilitary units. Examining Central and Southern California, McWilliams reports that "in the Salinas strike, 1,500 men were mobilized for deputy duty in less than a day; in the Stockton strike, 2,200 deputies were mobilized in a few hours and in Imperial Valley 1,200 deputies were recently mobilized on a few minutes' notice."[39] Through localized memberships, the corporate interests backing AF could channel their will through enormous vigilante groups, rivaling organized labor's numerical strength and threatening physical violence on top of that.

AF formed connections with a sprawl of civic associations and local militias, above- board and otherwise. Researchers Nelson Pichardo and Brian Kulik count "the United for California League, The Southern Californians, Inc., The Neutral Thousands, the California Citizens Association, the Committee of 43, The Women of the Pacific, the California Committee for Peace in Industrial Relations, the Agricultural Producers Labor Committee, and the California Cavaliers" among its peers and relatives, with the Chamber of Commerce and the Merchants and Machinery Association as their chief progenitors.[40]

The AF's local chapters and connections multiplied the organization's influence, allowing it to take root in rural California. Through AF, a disparate patchwork of civic associations and vigilantes were absorbed into a statewide antilabor infrastructure. This proliferation-by-proxy also created the impression that the antilabor cause which the AF championed represented an organic, widespread sentiment. The cluster of organizations plied public opinion, advocated for anti-labor legislation, and resorted to force when necessary.

33 *Violations of Free Speech and Rights of Labor Report*, 46.
34 *Violations of Free Speech and Rights of Labor Report*, 46.
35 McWilliams, *Factories in the Field*, 230.
36 McWilliams, *Factories in the Field*, 232.
37 *Violations of Free Speech and Rights of Labor Report*, 46.
38 *Violations of Free Speech and Rights of Labor Report*, 47.
39 McWilliams, *Factories in the Field*, 235.
40 Pichardo and Kulik, *American Fascism and the New Deal*, 114.

AF frequently made common cause with overtly reactionary organizations. For instance, the organization often enlisted American Legion chapters to aid in their strikebreaking endeavors. The American Legion, a veterans' association founded in 1919, had terrorized labor since its inception. On November 11 of 1919, Legionnaires stormed a "Wobbly" (a term referring to members of the Industrial Workers of the World, or IWW) hall in Centralia, Washington. They came bearing firearms, and began physically attacking Wobblies present. The latter group fired on the Legionnaires in self defense, killing three before succumbing to beating and mass arrest. That night, Legionnaires dragged one Wobbly, Wesley Everest, from the Centralia jail. Everest was beaten, castrated, hanged, and shot. Vigilantes dumped his corpse in front of the cells for the rest of the Wobblies to see.[41] If these actions were not enough, the American Legion voiced its support for fascism abroad. In 1923, for instance, Legionnaires invited Benito Mussolini to their annual convention in San Francisco.[42]

As previously mentioned, Legionnaires also assisted in the Sonoma tarring and feathering of Jack Green and Sol Nitzberg. In 1933, grape growers in Lodi, California mobilized local Legionnaires against striking farmworkers.[43] The American Legion was used to similar effect in the Imperial Valley in 1934, where they forcibly evicted two thousand striking farmworkers and indirectly killed an infant with tear gas.[44] The Imperial Valley Legionnaires' leader, Colonel Walter E. Garrison, went on to become one of the foremost militia organizers for AF.[45] Another AF militia organizer, Guernsey Frazier, also served as an official for the American Legion.[46] Harper Knowles, who conducted espionage on behalf of the AF, also headed the American Legion Radical Research Committee.[47] An exposé on the AF featured in the *Rural Observer* counted Garrison, Frazier, and Knowles among the organization's "Front Boys," mid-level officials who saw to the unsavory work of espionage and recruitment.[48] AF, the American Legionnaires, and the wider world of strikebreaking in California were all inextricably linked.

The anti-labor network made common cause with even more overtly fascistic organizations. In 1935, AF official Herman Cottrell founded the California Cavaliers in Sacramento. Cavaliers were usually deputized in Legionnaire halls, indicating a great degree of overlap between the Cavaliers, Legionnaires, and AF. According to researchers Pichardo and Kulik, "little is available on the activities of the Cavaliers, but they had been described as fascist along the same lines as the Silver Shirts."[49] The Silver Shirts, of course, were among the crowd who terrorized Green and Nitzberg. They espoused an explicitly antisemitic and pro-Nazi worldview. Founded in North Carolina in 1933, the organization spread and rose to particular prominence along the West Coast, with an outsized presence San Francisco.[50] Nitzberg explains that prior to the tarring and feathering, "We [the Communist Party] challenged the Silver Shirts to a debate at a mass meeting in

41 Michael Cohen, "'The Ku Klux Government': Vigilantism, Lynching, and the Repression of the IWW," *Journal for the Study of Radicalism* 1, no. 1 (2007): 48. http://www.jstor.org/stable/41887562.

42 "Legion Invites Mussolini," *The New Majority* (Chicago) August 4, 1923, Marxists.org.

43 Pichardo and Kulik, *American Fascism and the New Deal*, 69.

44 McWilliams, *Factories in the Field*, 225.

45 "Who Are the Associated Farmers?" *The Rural Observer* (San Francisco), 1937, Online Archive of California.

46 Pichardo and Kulik, *American Fascism and the New Deal*, 108.

47 Judkins, "Under Prying Eyes," 113.

48 "Who are the Associated Farmers?"

49 Pichardo and Kulik, *American Fascism and the New Deal*, 116.

Santa Rosa, and they turned out in the audience. We talked about the contributions the Jews have made to the morality of the world, to literature. We defended American democracy against these alien fascist ideas."[51] The Silver Shirts failed to make a counterargument, instead absconding from the meeting hall and taking their revenge on Nitzberg at a later date. While the Cavaliers and Silver Shirts are a minor presence in the antilabor landscape compared to the Legionnaires and AF, they represent the extremities of the network. The various organizations constituting California's antilabor network, AF chief among them, were willing to mobilize a reactionary paramilitary element if it paralyzed labor.

These reactionary sentiments were not limited to the vigilantes, either. Members of the AF leadership expressed admiration for Nazi Germany. In 1936, California Senator and AF member John Phillips praised Adolph Hitler and spoke about the German agricultural system in glowing terms.[52] On the subject of German citizenship, Phillips beamed that "you simply say that anyone who agrees with you is a citizen of the first class, and anybody who does not agree with you is a non-voting citizen."[53] Another AF leader, Philip Bancroft, asked John Phillips to reiterate his praise for German agriculture at a Fruit Growers and Farmers Convention, implying widespread approval for fascistic governing policies across Californian employer associations and strikebreaking organizations.[54]

Journalist Carey McWilliams goes so far as to suggest the existence of concentration camps in California. In *Factories in the Field* he writes of one in Salinas, claiming that "a stockade has been constructed which is admittedly intended for use as a concentration camp. When local workers inquired of the shipper-growers why such a curious construction had been made, they were told that it was built 'to hold strikers, but of course we won't put white men in it, just Filipinos.'" Similar camps cropped up in Brentwood and Tagus Ranch. In all three cases, employers crammed workers (usually strikers) into an enclosed area surrounded by walls, barbed wire, and armed guards. McWilliams likens these to "fortified camps."[55]

Employers and corporations founded a sprawl of antilabor organizations, which in turn made a concerted attempt to multiply across the state of California. These organizations connected with civic associations at the local level, and a network began to take shape. That network was willing to shape laws, threaten violence, and make common cause with fascist elements in order to suppress labor. Thus, the largest city and the smallest town responded to labor in identical ways, and the vigilante and the employer found themselves linked by a continuous thread.

## Looking North

A survey of antilabor vigilantism in Northern California reveals that the techniques and network pioneered in Southern and Central California carried over seamlessly, affecting organized labor across industry, transportation, and agriculture. The "fortified camps" in Brentwood and Tagus appeared in northern agriculture as well. Writing

50 Raymond Dannenbaum, "Fantasy in Uniform," *The Jewish News of Northern California* (San Francisco), March 9, 1934, California Digital Newspaper Collection.
51 Kann, *Comrades and Chicken Ranchers*, 104
52 "Who Are the Associated Farmers?"
53 McWilliams, *Factories in the Field*, 234.
54 "Who Are the Associated Farmers?"
55 McWilliams, *Factories in the Field*, 237-238.

in 1938, journalist Carey McWilliams reported that "A former relief worker has told how some four thousand men were taken off the relief rolls in Northern California. These men were corralled in 'concentration camps' described as 'pest holes without tents," with open latrines and a total absence of sanitation. Hundreds of cases of dysentery, hop poisoning (a skin disease), and influenza were reported in the camps."[56]

The basic units of vigilantism, the militias and citizen committees, developed across Northern California. These organizations maintained an exceedingly militant face, even when labor was on its back foot. In August of 1935 for instance, an article in the Sotoyome Scimitar newspaper reported that citizens of Napa County formed one such committee to "cope with a threatened invasion of that county by Communist agitators. With the alleged Communists, many of them driven from Sonoma County by the vigilantes, arriving and continued threats to proceed with their plans thwarted in Sonoma County, the Napa officers are preparing to adopt stringent measures to insure an orderly harvest."[57] The terms "threatened invasion" and "agitators" cast the communists as aggressive and unlawful. In actuality, the article was referring to labor organizers who had been kidnapped and beaten alongside Green and Nitzberg. They left Sonoma County under threat of violence after attempting to organize Sebastopol's apple orchard workers.[58] Through a rhetorical sleight of hand, the *Sotoyome Scimitar* recast the exercise of free speech and assembly as sinister.

A similar vigilante committee sprung up in Mill Valley. The city's Chamber of Commerce called an emergency meeting in July of 1934, in the wake of the Waterfront Strike gripping nearby San Francisco. At that meeting, Chief of Police Alex McCurdy claimed that local taxi drivers "had gone off duty without adequate reason" and subsequently approved a citizen's committee to ensure "law and order."[59] Local papers did not expound on this comment, but it is possible that Mill Valley taxi drivers had engaged in a wildcat strike in solidarity with the waterfront workers after Bloody Thursday, when San Francisco police killed two strikers and injured over one hundred over the course of a day-long clash.[60]

Although "strike leaders in San Francisco gave repeated assurance that strike activities [did] not extend to [Marin County]," community leaders conducted themselves as though they were under siege. A "Merchant Committee" stockpiled food and fuel, even as the San Francisco Teamsters stressed that Mill Valley had no cause for concern.[61] Meanwhile, the *Mill Valley Record* reported that the citizen's committee had deputized one hundred citizens, who established checkpoints and barricades around the town. The paper boasted that "the precautions taken by Mill Valley Citizen's Committee to prevent Communist activity or reprisals have assured safety and order to the citizens and business of the town."[62] These actions were an incongruous and disproportionate response to supposedly recalcitrant taxi drivers and a nearby strike whose leaders were wholly disinterested in Mill Valley. The checkpoints and

56 McWilliams, *Factories in the Field*, 288.

57 "Napa Residents Plan to Form Vigilant Body," *Sotoyome Scimitar* (Healdsburg), August 22, 1935, California Digital Newspaper Collection.

58 "Defiant Reds are Driven Out of County in Sporadic Raid," *Sotoyome Scimitar* (Healdsburg), August 22, 1935, California Digital Newspaper Collection.

59 "Citizens Form Vigilante Night Patrol Guards," *Mill Valley Record*, July 20, 1934, California Digital Newspaper Collection.

60 David F. Selvin, "An Exercise in Hysteria: San Francisco's Red Raids of 1934," *Pacific Historical Review* 58, no. 3 (1989): 361. https://doi.org/10.2307/3640271.

61 "Citizens Form Vigilante Night Patrol Guard."

stockpiling are better understood as a message: any sign of labor unrest in Mill Valley would be construed as communist infiltration, which would warrant suppression and intimidation.

Intimidation frequently gave way to violence. In September of 1933, in Chico, the Journey family found a letter inside of their car. It read as follows: "Get your bunch out of the county or you will be put on the spot. The bottom of Feather River will be your last move. Remember, life is sweet. You have caused me enough trouble for one year."[63] The family's son, James Journey, had helped lead a strike at the Steadman Peach Ranch in nearby Butte County in the previous month. Moreover, he was set to testify against the ranch owner's son, Ernest Steadman. In August, Ernest Steadman drove a truck into the picket line and hit two strikers, after which point a local chapter of American Legionnaires descended on the scene to "prevent a serious outbreak" of violence.[64] The ILD was pursuing a damage suit of $100,000 against Steadman, with James Journey as a key witness.[65] When the Journey family chose to stay put, five masked men took James Journey at gunpoint. They beat him for several days, then stranded him in Colusa County.[66] The Chico kidnapping mirrors high profile episodes such as an abduction at Pixley (wherein vigilantes kidnapped and threatened A.L. Wirin, an ACLU lawyer representing striking cotton workers in the Imperial Valley) and the tarring-and-feathering in Sonoma County.[67] In all three cases, vigilantes singled out leaders and key figures to terrorize and hinder the broader labor movement.

Vigilantes also carried out mass expulsions in Northern California. In April of 1938, vigilantes ousted between one and two hundred (estimates vary in press coverage of the incident) CIO-affiliated miners and their families from the city of Grass Valley, in Nevada County. The workers had struck against Murchie Mine in the weeks leading up to the expulsion.[68] When individual strikers returned to gather their belongings, at least three reported threats. A few arrived under the guard of California State Highway Patrol.[69] Executive Director of the Northern California ACLU Ernest Besig expressed horror at the episode, understanding it as a total upheaval of civil liberties.[70] One *Imperial Valley Press* article dubbed the situation a "choice between union and citizen," wherein the union represented "outside agitators" steeped in the "Communist book of propaganda," while citizens possessed an "inborn respect for the Constitution."[71] Others took the allegations of vigilantism in stride. Editor Charles Hastings Deuel wrote that:

> It was action not according to law and statutes, but it was effective...The actions of the citizens of Grass Valley and Nevada City will be widely denounced. There will be allegations of the violation of civil rights, of violence against personal liberty. There always is when 'beef squads' are in turn subjected to their own methods. The

---

62 "Citizens Form Vigilante Night Patrol Guard."

63 "Young Chico Strike Leader Kidnapped, Held Two Days," *The Western Worker* (San Francisco), October 9, 1933, California Digital Newspaper Collection.

64 "Two Strikers Run Down in Gridley Area," *Imperial Valley Press*, August 23, 1933, California Digital Newspaper Collection.

65 "Young Chico Strike Leader."

66 "Agricultural Workers Attacked," *Labor Defender* (New York), November, 1933, Marxists.org.

67 Taylor, *On the Ground in the Thirties*, 25.

68 "Murchie Mine Union Disturbance," Reno-Gazette Journal, May 19, 1938, Newspapers.com.

69 "Grass Valley Strike Group Returns Under Guard to Mine Area," *Imperial Valley Press*, April 13, 1938, California Digital Newspaper Collection.

70 "Vigilantes Rule Nevada County," *ACLU of Northern California News* (San Francisco), May, 1938, California Historical Society.

71 "The Choice Between Union and Citizens," *Imperial Valley Press*, April 13, 1938, California Digital Newspaper Collection.

> incidents occurring at these mountain communities, unlawful as they no doubt were, may operate to promote peace in other parts of rural California.[72]

Grass Valley set the tenor. Other expulsions would follow.

The episode repeated in Lassen County some months later. On July 8 of 1938, CIO lumber workers in the town of Westwood moved to strike against the Red River Lumber Company after the latter had imposed a seventeen percent wage cut. Prior to this, the lumber workers had been split between the company-dominated Industrial Employees' Union (IEU) and the CIO-affiliated Lumber Workers of America. According to an investigation conducted by the American Civil Liberties Union, some three hundred workers switched from the IEU to the CIO union during the strike.[73] Meanwhile, Lassen County Sheriff Olin Johnson deputized several hundred workers from the IEU.[74]

On July 13 of 1938, the deputized force descended on the CIO picket line bearing guns, pickaxes, and firehoses. One striker later claimed that the Red River Lumber Company supplied the weapons to the vigilantes. The *Calexico Chronicle* reported thirty injuries from the attack on both sides, but provided no further breakdown.[75] The vigilantes vandalized the CIO soup kitchen, menaced sympathizers, and drove eight hundred strikers out of Westwood. Strikers and their families scattered to the winds, though some two hundred and fifty arrived in Sacramento to plead their case to California Governor Frank Merriam.[76] When the California Highway Patrol escorted an investigative body into Westwood, crowds mocked and menaced them, and a string of cars trailed after the committee as a tacit threat.

The breakdown and subsequent suppression of a lumber strike at the Holmes-Eureka mill in Humboldt County showcased common vigilante tactics and rhetoric. On June 21, 1935, police launched tear gas and opened fire on a picket line, killing three strikers and wounding others at the Holmes-Eureka Mill.[77] These picketers had been on the fringes of the 1935 Pacific Northwest lumber strike (mostly concentrated in Washington and Oregon), and had fought an uphill battle since May of that year, as they possessed far less bargaining power than their counterparts in Oregon.[78] Shortly after the police opened fire, vigilantes swarmed the city of Eureka. They blocked off roads leading out of the city, vandalized a striker shantytown, and raided the town's Finnish Federation Hall (claiming that some toy weapons used for local theater performances constituted a weapons cache for communist radicals).[79]

When LaFollette Committee investigators began conducting preliminary hearings in 1936, Humboldt strikers reached out to them. The final Senate Report omitted their testimony entirely—perhaps for brevity's sake, or due to the time crunch and lack of funding which characterized the committee's time in California. The only record of the hearings comes Clara St. Peter, a local writer and schoolteacher married to one of the strikers. That

---

72 Charles Deuel, "Grass Valley Agitators Run Out by Miners," *Imperial Valley Press*, June 5, 1938, California Digital Newspaper Collection.

73 "A Report on California's Latest Vigilante Outrage," *ACLU of Northern California News* (San Francisco), August, 1938, California Historical Society.

74 "A Report on California's Latest Vigilante Outrage."

75 "Week at the State Capitol," *Calexico Chronicle* (El Centro), July 28, 1938, California Digital Newspaper Collection.

76 "A Report on California's Latest Vigilante Outrage."

77 Frank Onstine, *The Great Humboldt Lumber Strike 1935* (California: Lychgate Press, 2012), 490-483. Kindle.

78 Onstine, *The Great Humboldt Lumber Strike*, 37.

79 Onstine, *The Great Humboldt Lumber Strike*, 490.

said, the hearings revealed a great deal of overlap between the police, the vigilantes, the lumber mills, and a key member of California's anti-labor network.[80]

One former Dolbeer-Carson Lumber Mill employee John McNichols testified that vigilantes had coalesced under a militia known as the Humboldt Nationals, with the financial support and direct involvement of local lumber mills. He claimed that the mill paid him for the time that he spent drilling with the militia (using clubs manufactured at the nearby Hammond Mill). Additionally, a company lawyer attended these militia meetings. Local police called on the Humboldt Nationals several times throughout the strike, and a sheriff deputized some two hundred and fifty members into the organization.[81] The militia was organized by one Colonel Sanborn, an out-of-towner who managed a right-wing newspaper based in San Rafael. A local observer reported that: "It was learned that organizing such 'law-and-order' bodies as the Humboldt Nationals was a professional occupation of the colonel, he having started such organizations in different parts of the state."[82]

Colonel Sanborn's presence is noteworthy. He was a consistent fixture in the AF network. The Rural Observer exposé on the organization counted him among its "Front Boys," men who conducted espionage and organized militias to undermine labor.[83] He was a reserve officer in a military intelligence unit. The ACLU press harbored a particular distaste for these middle men, and singled out Sanborn as a "red-baiter" across multiple issues.[84] Besides Eureka, Sanborn had organized militias in Salinas, Crockett, Berkeley, Santa Cruz, and San Benito. Sanborn used red- baiting as a cudgel against labor. During the 1934 Salinas Lettuce Strike, he claimed that red flags planted along a highway in Salinas signaled a radical communist presence in the area. These flags were construction markers.[85] During this time, Sanborn also conducted an illegal search in a striker's home and seized communist literature, which he likely planted.[86]

Sanborn's aforementioned newspaper, *American Citizen*, invoked the following masthead: "In order that fascism may not become necessary to prevent communism from becoming a reality."[87] A 1936 issue of *ACLU of Northern California News* pointed out that the paper counted AF and the Waterfront Employers Association among its chief beneficiaries.[88] American Citizen had very few subscribers, and likely relied on these moneyed organizations to stay afloat. Associated Farmer meeting minutes collected by the LaFollette Committee include tables for their expenses. In 1936 alone, the organization purchased at least 3,500 copies of *American Citizen* papers, with the "Payee" alternating between Sanborn and the publication itself.[89]

Sanborn was a jack-of-all-trades for the AF, and emblematic of the "front" set. More than this, however, his presence highlighted just how

80 Onstine, *The Great Humboldt Lumber Strike*, 870.
81 Onstine, *The Great Humboldt Lumber Strike*, 876.
82 Onstine, *The Great Humboldt Lumber Strike*, 890.
83 "Who Are the Associated Farmers?"
84 "Colonel Henry Roble Sanborn; Unfrocked, Unspeakable, Unashamed," ACLU of Northern California News, November, 1936, California Digital Newspaper Collection; "The Renascence of Col. Henry Sanborn, Alias, 'Mr. Winters,'" *ACLU of Northern California News*, November, 1940, California Digital Newspaper Collection.
85 "Colonel Henry Roble Sanborn; Unfrocked, Unspeakable, Unashamed."
86 "Colonel Henry Roble Sanborn; Unfrocked, Unspeakable, Unashamed."
87 "The Renascence of Col. Henry Sanborn, Alias, 'Mr. Winters.'"
88 "Colonel Henry Roble Sanborn; Unfrocked, Unspeakable, Unashamed."
89 Violations of Free Speech and Rights of Labor: Hearings before a *Subcommittee of the Committee on Education and Labor, United States Senate, Seventy-Sixth Congress, Third Session, Pursuant to S. Res. 266* (Washington, D.C., United States Government Printing Service, 1940), 25,697-25,698.

pervasive AF and the antilabor network really was. One of its front men had interfered in an industry completely separate from agriculture or even shipping, using tactics pioneered in the Imperial Valley and the board rooms of Remington Rand.

## Conclusion: The Choice Between Union and Citizens

All of the patterns of strikebreaking associated with Central and Southern California were replicated in the North. The tactics were identical, sometimes with the same names and faces cropping up. Through the pretense of a citizens' committee, the will of the employer became part and parcel with a right wing populist construction of community. Lines were drawn and workers were shut out of this constructed community, even if they were the majority of its population and made it viable to begin with. The outsider became the enemy, and the enemy was brutalized by the employer's coalition of law enforcement and vigilantes. Anti-labor forces paradoxically disregarded the law and preached adherence to it— some operated with a patina of legitimacy, within deputized paramilitary units, but willing to resort to kidnapping and extralegal violence whenever necessary. These vigilante tactics predate the 1930s. They have their roots in the state's early history. By the 1930s, however, that parochial vigilantism had the full financial backing and coordination of such moneyed organizations as the Chamber of Commerce. In fact, those institutions became the generative force of vigilantism. An able set of ex-military men moved across the state of California, replicating militias and nudging public opinion wherever labor was beginning to coalesce. The men who cut their teeth in the military (including Henry Sanborn and Walter E. Garrison, to name some of the aforementioned "Front Boys") brought the war home, and labor found itself on the other side of the line. These militias constituted an extremist element, and made common cause with even more overtly extremist elements.

Strikers and labor organizers, viewed as separate from the citizenry, had to fight an uphill battle to exercise their most basic civil liberties. They could be misrepresented in the press, driven from their homes, brutalized on the picket line, or menaced by vigilantes, with little recourse. They were made to live alongside their perpetrators. In this way, California's antilabor network dealt incalculable harm to organized labor and the social fabric.

## Bibliography

"Agricultural Workers Attacked." *Labor Defender* (New York), November, 1933. Marxists.org.

"A Report on California's Latest Vigilante Outrage." *ACLU of Northern California News* (San Francisco), August, 1938. California Historical Society.

Auerbach, Jerold S. "The La Follette Committee: Labor and Civil Liberties in the New Deal." *The Journal of American History* 51, no. 3 (1964): 435–59. https://doi.org/10.2307/1894895.

Chacón, Justin Akers and Davis, Mike. *No One is Illegal: Fighting Violence and State-Repression on the U.S.-Mexico Border.* Chicago, Illinois: Haymarket Books, 2006.

"Citizens Form Vigilante Night Patrol Guards." *Mill Valley Record*, July 20, 1934. California Digital Newspaper Collection.

Cohen, Michael. "'The Ku Klux Government': Vigilantism, Lynching, and the Repression of the IWW." *Journal for the Study of Radicalism* 1, no. 1 (2007): 31–56. http://www.jstor.org/stable/41887562.

"Colonel Henry Roble Sanborn; Unfrocked, Unspeakable, Unashamed." *ACLU of Northern California News*, November, 1936. California Digital Newspaper Collection.

Dannenbaum, Raymond. "Fantasy in Uniform." *The Jewish News of Northern California* (San Francisco), March 9, 1934. California Digital Newspaper Collection.

"Defiant Reds are Driven Out of County in Sporadic Raid." *Sotoyome Scimitar* (Healdsburg), August 22, 1935. California Digital Newspaper Collection.

Deuel, Charles. "Grass Valley Agitators Run Out by Miners." *Imperial Valley Press*, June 5, 1938. California Digital Newspaper Collection.

Fronczak, Joseph. "The Fascist Game: Transnational Political Transmission and the Genesis of the U.S. Modern Right." *The Journal of American History* 105, no. 3 (2018): 563–88. https://www.jstor.org/stable/48549492.

"Grass Valley Strike Group Returns Under Guard to Mine Area." *Imperial Valley Press*. April 13, 1938.

Janiewski, Dolores E. "Through a Glass, Darkly: The NLRB, Employer Counter-offensives, Investigative Committees, and the CIO." In *Against Labor: How U.S. Employers Organized to Defeat Union Activism*. Champaign, Illinois: University of Illinois Press, 2017. http://www.jstor.org/stable/10.5406/j.ctt1p6qq3f.9.

Judkins, Simon J. "Under Prying Eyes: Repression, Surveillance, and Exposure in California, 1918-1939." Master's thesis. Victoria University of Wellington, 2014.

Kann, Kenneth L. *Comrades and Chicken Ranchers: The Story of a California Jewish Community*. Ithaca: Cornell University Press, 1993.

Kulik, Brian W. and Pichardo, Neson A. *American Fascism and the New Deal: The Associated Farmers of California and the Pro-Industrial Movement.* Lanham: Lexington Books, 2013.

"Legion Invites Mussolini." *The New Majority* (Chicago), August 4, 1923. Marxists.org.

McWilliams, Carey. *Factories in the Field.* Boston: Little, Brown and Company, 1939.

"Murchie Mine Union Disturbance." *Reno-Gazette Journal*, May 19, 1938. Newspapers.com.

"Napa Residents Plan to Form Vigilante Body." *Sotoyome Scimitar* (Healdsburg), August 22, 1935.

Onstine, Frank. *The Great Humboldt Lumber Strike 1935*. California: Lychgate Press, 2012. Kindle.

"Ousted Westwood Workers Protest." *Santa Ana Journal*, July 15, 1938.

Pisani, Donald J. "Land Monopoly in Nineteenth-Century California." *Agricultural History* 65, no. 4 (1991): 15–37. http://www.jstor.org/stable/3743944.

Ross, Philip and Taft, Philip. "American Labor Violence: Its Causes, Character, and Outcome." *The History of Violence in America: A Report to the National Commission on the Causes and Prevention of Violence.* ed. Hugh Davis Graham and Ted Robert Gurr. 1969.

Selvin, David F. "An Exercise in Hysteria: San Francisco's Red Raids of 1934." *Pacific Historical Review* 58, no. 3 (1989): 361–74. https://doi.org/10.2307/3640271.

Taylor, Paul. *On the Ground in the Thirties*. Layton: Peregrine Smith Books, 1983.

"The Choice Between Union and Citizens." *Imperial Valley Press*, April 13, 1938. California Digital Newspaper Collection.

"The Renascence of Col. Henry Sanborn, Alias, 'Mr. Winters.'" *ACLU of Northern California News*, November, 1940. California Digital Newspaper Collection.

"Two Strikers Run Down in Gridley Area." *Imperial Valley Press*, August 23, 1933. California Digital Newspaper Collection.

"Vigilante Night Patrol Guards." *Mill Valley Record*, July 20, 1934.

"Vigilantes Rule Nevada County." *ACLU of Northern California News* (San Francisco), May, 1938. California Historical Society.

*Violations of Free Speech and Rights of Labor. Report of the Committee on Education and Labor, Pursuant to S. Res. 266 Seventy-Fourth Congress, a Resolution to Investigate Violations of the Right of Free Speech and Assembly and Interference With the Right of Labor*.

Washington D.C.: United States Government Printing Office. 1942.

*Violations of Free Speech and Rights of Labor: Hearings before a Subcommittee of the Committee on Education and Labor, United States Senate, Seventy-Sixth Congress, Third Session, Pursuant to S. Res. 266*. Washington, D.C., United States Government Printing Service. 1940.

"Week at the State Capitol." *Calexico Chronicle* (El Centro), July 28, 1938. California Digital Newspaper Collection.

White, Ahmed A. "Industrial Terrorism and the Unmaking of the New Deal." *Nevada Law Journal* 11, no. 561 (2011): 561-628.

"Who Are the Associated Farmers?" *The Rural Observer,* 1937. Online Archive of California.

"Young Chico Strike Leader Kidnapped, Held Two Days." *The Western Worker* (San Francisco), October 9, 1933. California Digital Newspaper Collection.

Made in the USA
Monee, IL
02 November 2026

67376799R00057